Y0-BYE-506

Bilkent University Lecture Series

Sanjay Ranka
Sartaj Sahni

Hypercube Algorithms

with Applications to Image Processing and Pattern Recognition

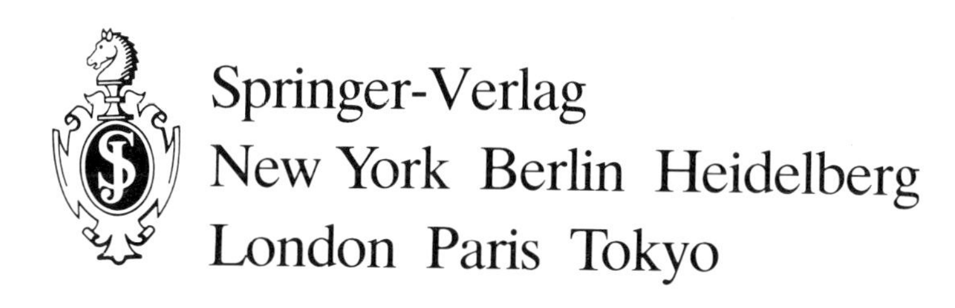

Springer-Verlag
New York Berlin Heidelberg
London Paris Tokyo

Prof. SANJAY RANKA
School of Computer and Information Science
Syracuse University
Syracuse, NY 13090
U.S.A.

Prof. SARTAJ SAHNI
Computer Science Department
University of Minnesota
Minneapolis, Minnesota 55455
U.S.A.

Library of Congress Cataloging-in-Publication Data
Ranka, Sanjay,
Hypercube algorithms: with applications to image processing and pattern recognition/
Sanjay Ranka, Sartaj Sahni.
p. cm.
Includes bibliographical references.
ISBN 0-387-97322-2 (U.S.)
1. Parallel processing (Electronic computers). 2. Algorithms. 3. Hypercube networks (Computer networks). 4. Image processing-Digital techniques. 5. Pattern recognition systems. I. Sahni, Sartaj. II. Title.
QA76.58.R36 1990 90-9698
006.4--dc20 CIP

Printed and bound by Meteksan A.Ş., Ankara, Turkey.
Printed on acid-free paper produced by Meteksan A.Ş.

9 8 7 6 5 4 3 2 1

ISBN 0-387-97322-2 Springer-Verlag New York Berlin Heidelberg
ISBN 3-540-97322-2 Springer-Verlag Berlin Heidelberg New York

ISBN 975-7679-02-X Bilkent University Ankara

Preface

This book deals primarily with algorithmic techniques for SIMD and MIMD hypercubes. These techniques are described in detail in Chapter 2 and then used in subsequent chapters. Problems with application to image processing and pattern recognition are used to illustrate the use of the primitive hypercube operations developed in Chapter 2. The primitive operations of Chapter 2, however, have application to problems from other application areas too. Computational geometry, graph theory, scheduling, and VLSI CAD, are examples of some of the other application areas where these techniques have been applied. In addition to dealing with techniques for fine grained hypercubes, techniques for medium grained hypercubes such as those currently available are also discussed.

This book is suitable for use in a one semester or one quarter course on hypercube algorithms. For students with no prior exposure to parallel algorithms, it is recommended that one week be spent on the material in Chapter 1, approximately six weeks on the material in Chapter 2, and one week on Chapter 3. The remainder of the term can be spent covering selected topics from the rest of the book.

Some of the subject matter covered in this book was presented in a ten week seminar taught at the University of Minnesota. A preliminary version of this book was used in an intensive two week seminar at

Bilkent University, Ankara. The authors are greatly indebted to the students of this seminar as well as to Professors Cevdet Aykanat and Fikret Ercal of Bilkent University for their many suggestions that have resulted in an improved manuscript.

The authors are also indebted to Professors Mehmet Baray, Ali Doğramacı, and Özay Oral, for their encouragement and support in the development of this book.

Sanjay Ranka
Syracuse University

Sartaj Sahni
University of Minnesota

Contents

Chapter 1

Introduction

1.1 Parallel Architectures

Parallel computers may be classified by taking into account their memory organization, processor organization, and the number of instruction streams supported.

Memory organization

A *multiprocessor* is a parallel computer that has two or more processors. These processors share a common memory or a common memory address space (Quinn 1987). A block diagram of a *tightly coupled* multiprocessor is provided in Figure 1.1. In such a computer, the processors access memory via a processor-memory interconnection network. This network could be a simple bus or any of a variety of switching networks such as the Omega network, Benes network, full cross bar switch, etc. (Siegel 1979). In a *loosely coupled* multiprocessor, each processor has a local memory (Figure 1.2). These local memories together form the shared address space of the computer. Typically a memory reference to the local memory of a processor is orders of magnitude faster than a memory reference to a remote memory as local memory references are not routed through the interconnection network while remote memory references are.

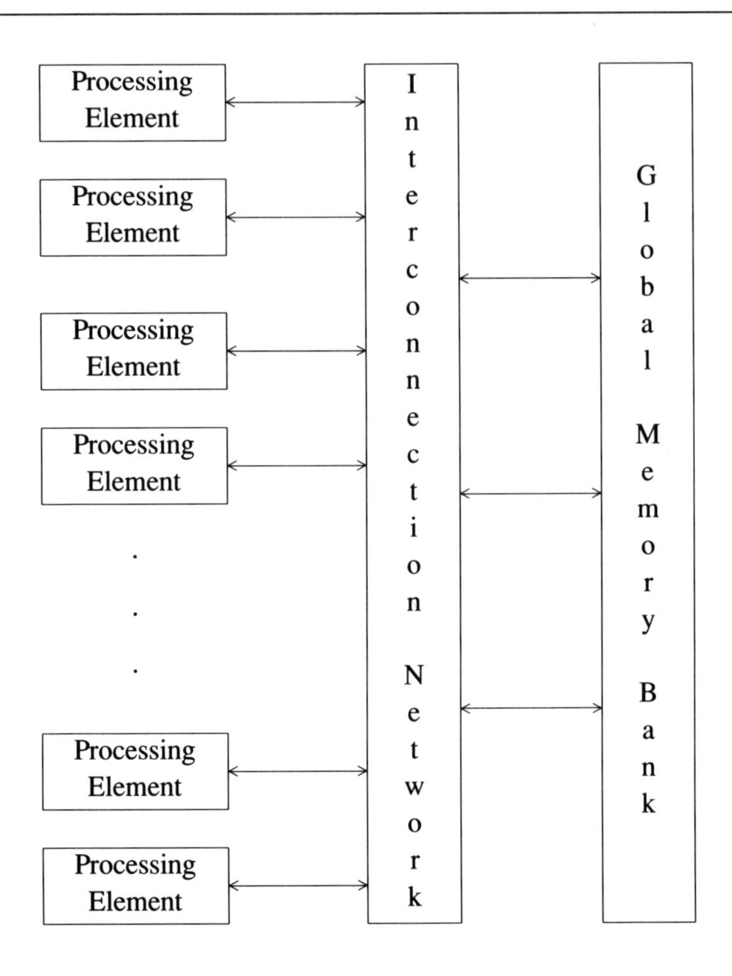

Figure 1.1 Tightly coupled multiprocessor

The block diagram for a *multicomputer* is the same as that for a loosely coupled multiprocessor (Figure 1.2). The significant difference between a multicomputer and a multiprocessor is that a multicomputer has neither a shared memory nor a shared memory address space (Quinn

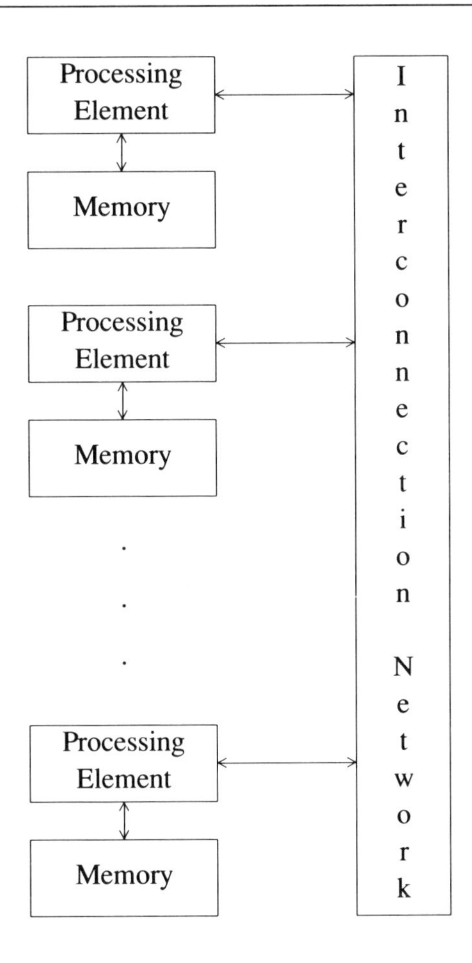

Figure 1.2 Block diagram for a loosely coupled multiprocessor and a multicomputer

1987). Consequently to use data in a remote memory, it is necessary to explicitly move that data into the local memory. This and all other interprocessor communication is done by passing messages (via the interconnection network) among the processors. Our further discussion is restricted to multicomputers.

Processor organization

Processor organization is defined by the interconnection network used to connect the processors of the multicomputer. Some of the more common interconnection networks are: two dimensional mesh, ring, tree, and hypercube (Figure 1.3).

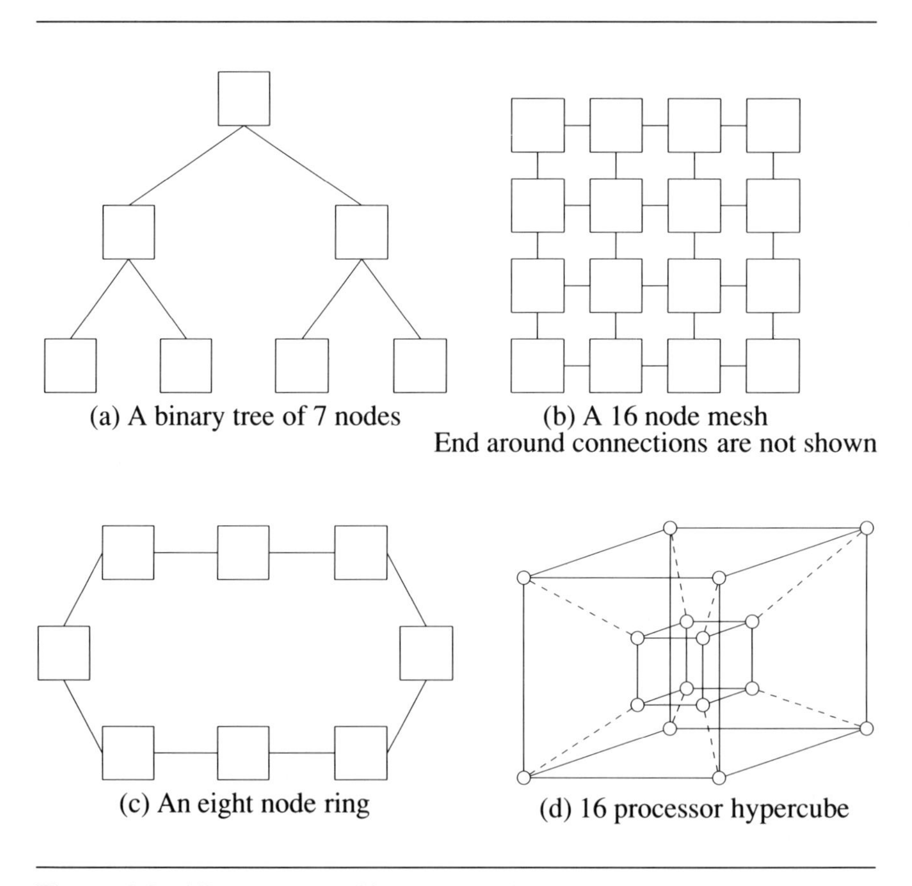

Figure 1.3 Different types of interconnection networks

The first three are intuitive while the fourth needs some elaboration. In a hypercube of dimension d, there are 2^d processors. Assume that these are labeled $0, 1, \cdots, 2^d - 1$. Two processors i and j are directly connected iff the binary representations of i and j differ in exactly one bit. Each edge of Figure 1.3 (d) represents a direct connection. Thus in a hypercube of dimension d, each processor is connected to d others. If the direct connection between a pair of processors i and j is *unidirectional*, then at any given time messages can flow from either i to j or from j to i. In the case of *bidirectional* connections, it is possible for i to send a message to j and for j to simultaneously send one to i.

The popularity of the hypercube network may be attributed to the following:

(1) Using d connections per processor, 2^d processors may be interconnected such that the maximum distance between any two processors is d. While meshes, rings, and binary trees use a smaller number of connections per processor, the maximum distance between processors is larger. It is interesting to note that other networks such as the star graph (Akers, Harel, and Krishnamurthy 1987) do better than a hypercube in this regard. A star graph connects $(d+1)!$ processors using d connections per processor. The inter-processor distance is at most $\left\lfloor \frac{3(d-1)}{2} \right\rfloor$. The hypercube has the advantage of being a well studied network while the star graph is relatively new and few algorithms have been developed for it.

(2) Most other popular networks are easily mapped into a hypercube. For example a 2×4 mesh, 8 node ring, and a 7 node full binary tree may be mapped into an 8 node hypercube as shown in Figure 1.4. We shall examine these mappings in detail later.

(3) A hypercube is completely symmetric. Every processor's interconnection pattern is like that of every other processor. Furthermore, a hypercube is completely decomposable into sub-hypercubes (i.e., hypercubes of smaller dimension). This property makes it relatively easy to implement recursive divide-and-conquer algorithms on the hypercube (Stout 1987).

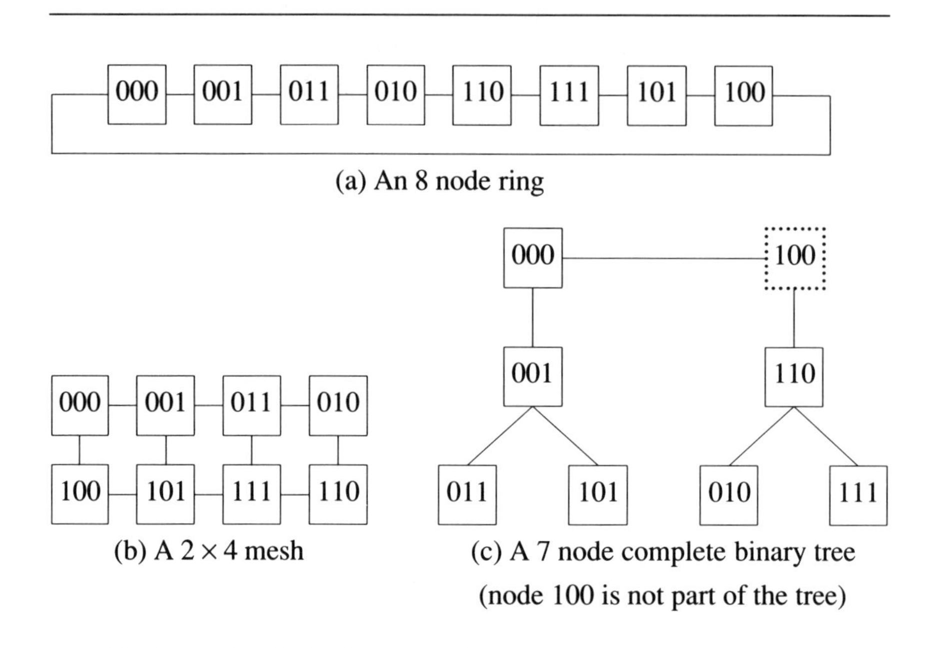

Figure 1.4 Embedding of different networks in an 8 node hypercube

Instruction streams

Flynn (1966) classified computers based on the number of instruction and data streams. The two categories relevant to our discussion here are SIMD (single instruction multiple data streams) and MIMD (multiple instruction multiple data streams) computers. In an SIMD parallel computer, all processors execute in a synchronous manner. In any given cycle, all processors execute the same instruction. MIMD parallel computers are generally asynchronous (in theory they could be synchronous too) and different processors may execute different instructions at any given time.

We shall consider both SIMD and MIMD multicomputers. Block diagrams of these are given in Figure 1.5 and Figure 1.6, respectively. In this book we are concerned only with the case when the interconnection network is the *hypercube network*. The important features of an SIMD hypercube computer and the programming notation we use are:

(1) There are $P = 2^p$ processing elements (PE). Each PE has a unique index in the range $[0, P-1]$. A p dimensional hypercube network connects the $P = 2^p$ PEs. Let $i_{p-1} i_{p-2} \cdots i_0$ be the binary representation of the PE index i. Let $\bar{i}_k$ be the complement of bit i_k. A hypercube network directly connects pairs of processors whose indices differ in exactly one bit. I.e., processor $i_{p-1} i_{p-2} \cdots i_0$ is connected to processors $i_{p-1} \cdots \bar{i}_k \cdots i_0$, $0 \le k \le p-1$.

(2) We use the notation $i^{(b)}$ to represent the number that differs from i in exactly bit b. We shall use brackets ([ ]) to index an array and parentheses ('( )') to index PEs. Thus A[i] refers to the i'th element of array A and A(i) refers to the A register of PE i. Also, A[j](i) refers to the j'th element of array A in PE i. The local memory in each PE holds data only (i.e., no executable instructions). Hence PEs need to be able to perform only the basic arithmetic operations (i.e., no instruction fetch or decode is needed).

(3) There is a separate program memory and control unit. The control unit performs instruction sequencing, fetching, and decoding. In addition, instructions and masks are broadcast by the control unit to the PEs for execution. An *instruction mask* is a boolean function used to select certain PEs to execute an instruction. For example, in the instruction

$$A(i) := A(i) + 1, \ \ (i_0 = 1)$$

$(i_0 = 1)$ is a mask that selects only those PEs whose index has bit 0 equal to 1. I.e., PEs with an odd index increment their A registers by 1. Sometimes we shall omit the PE indexing of registers. So, the above statement is equivalent to the statement:

$$A := A + 1, \ \ (i_0 = 1)$$

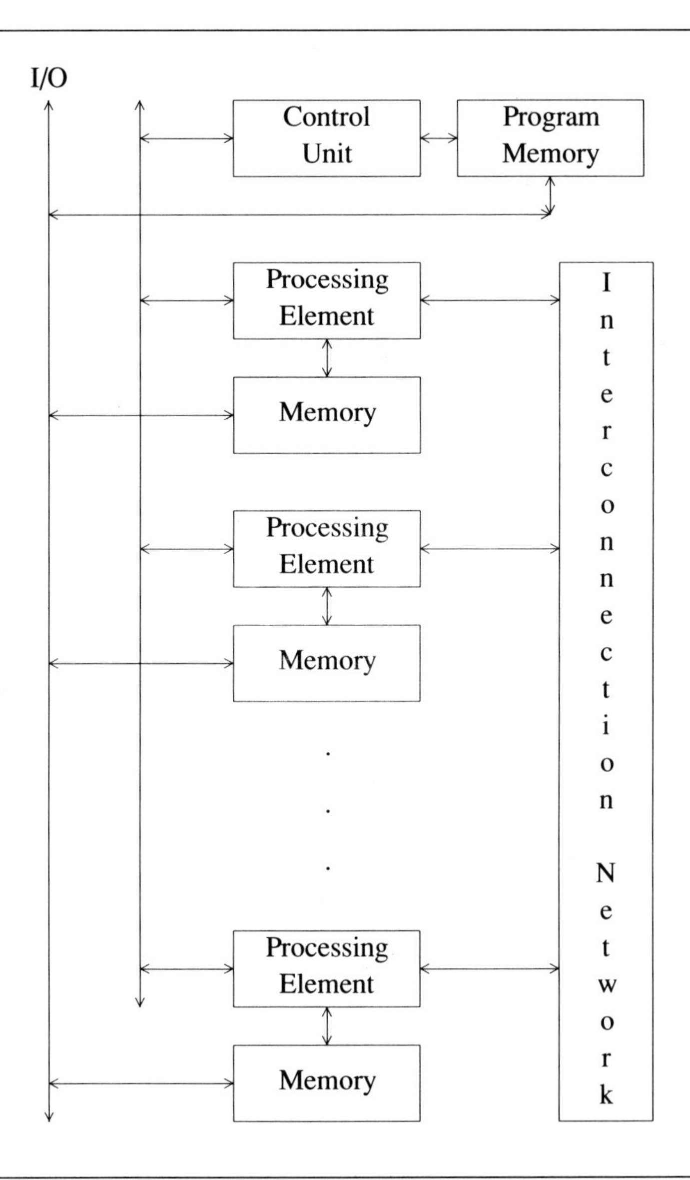

Figure 1.5 Block diagram of an SIMD multicomputer

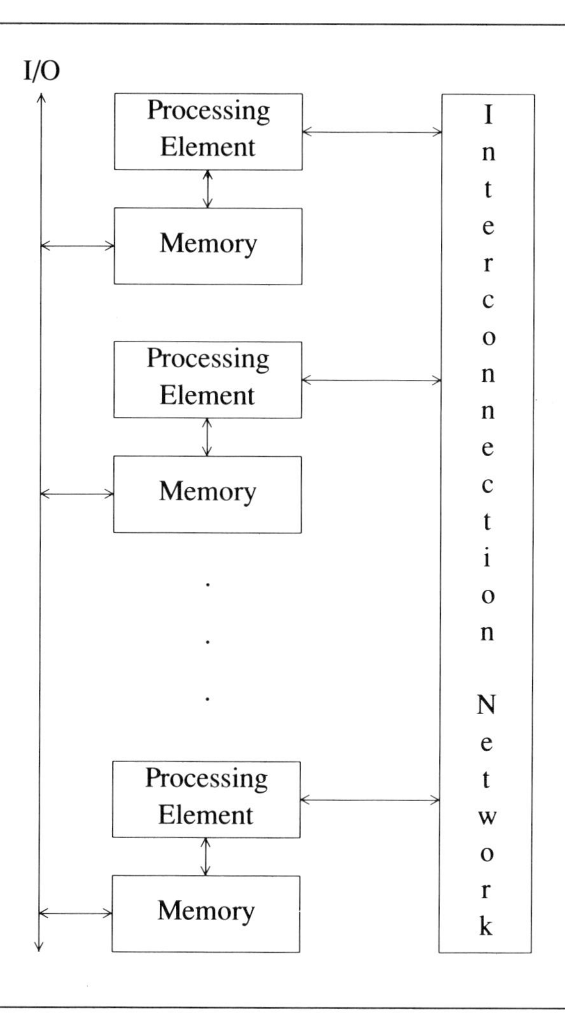

Figure 1.6 Block diagram of an MIMD multicomputer

(4) Interprocessor assignments are denoted using the symbol "←", while intraprocessor assignments are denoted using the symbol ":=". Thus the assignment statement:

$$B(i^{(2)}) \leftarrow B(i), \ \ (i_2 = 0)$$

on a hypercube is executed only by the processors with bit 2 equal to 0. These processors transmit their B register data to the corresponding processors with bit 2 equal to 1.

(5) In a *unit route*, one unit (bit, byte, word, or fixed size record) of data may be transmitted between pairs of processors that are directly connected. Each processor can send and/or receive only one unit of data in a unit route. If the links are unidirectional, then in a unit route data can be transferred only in one direction. If the links are bidirectional than data can be transferred in both directions in a unit route. For example, on a hypercube with unidirectional links, the instruction:

$$B(i^{(2)}) \leftarrow B(i), (i_2 = 0)$$

takes one unit route, while the instruction:

$$B(i^{(2)}) \leftarrow B(i)$$

takes two unit routes. If the hypercube has bidirectional links then both of the above instructions take one unit route each. Note also that in the case of an SIMD hypercube, different processors cannot transfer along different hypercube dimensions simultaneously. So, if one processor is transferring data to its neighbor along dimension 0, then another processor cannot simultaneously transfer data to its neighbor along dimension 1.

(6) Let B be the bandwidth of the links in the hypercube. The time required to transfer L units of data between two processors that are distance d apart is $O(dL/B)$ when the *store-and-forward* mechanism is used and $O(d + L/B)$ when *wormhole routing* is used (Athos and Seitz, 1988). Wormhole routing transmits the data in a pipelined manner using a sequence of L/B packets. The store-and-forward mechanism does not use pipelining. Note that when several processors are transmitting data, one needs to factor in the effects of possible path conflicts. The distinction between the store-and-forward and wormhole routing mechanisms isn't very important to the developments in this book as most of our data

transfers have $d = 1$ and involve very small amounts of data.

The features, notation, and assumptions for MIMD multicomputers differ from those of SIMD multicomputers in the following respects:

(1) There is no separate control unit and program memory.

(2) The local memory of each PE holds both the data and the program that the PE is to execute.

(3) At any given instance, different PEs may execute different instructions. In particular in an MIMD hypercube, PE i may transfer data to PE $i^{(b)}$, while PE j simultaneously transfers data to PE $j^{(a)}$, $i \neq j$ and $a \neq b$.

1.2 Embedding In A Hypercube

1.2.1 Definitions

A set of interconnected processors may be modeled as an undirected graph in which each vertex denotes a unique processor and there is an edge between two vertices iff the corresponding processors are directly connected. Let G and H be two undirected graphs that model two sets of interconnected processors. Let $V(G)$ and $V(H)$, respectively, denote the vertices in G and H. An *embedding* of G into H is a mapping of the vertices of G into the vertices of H and of the edges of G into simple paths of H. The vertex mapping is such that each vertex of G is mapped to a distinct vertex of H. Note that for an embedding to exist $|V(H)| \geq |V(G)|$. Also, if H is connected and $|V(H)| \geq |V(G)|$ then an embedding always exists.

As an example, consider the graphs G and H of Figure 1.7. The vertex mapping ($1 \rightarrow$ a, $2 \rightarrow$ b, $3 \rightarrow$ c, $4 \rightarrow$ d) and the edge to path mapping ((1,2) $\rightarrow$ ab, (2,4) $\rightarrow$ bad, (3,4) $\rightarrow$ cad, (1,3) $\rightarrow$ ac) defines an embedding of G into H.

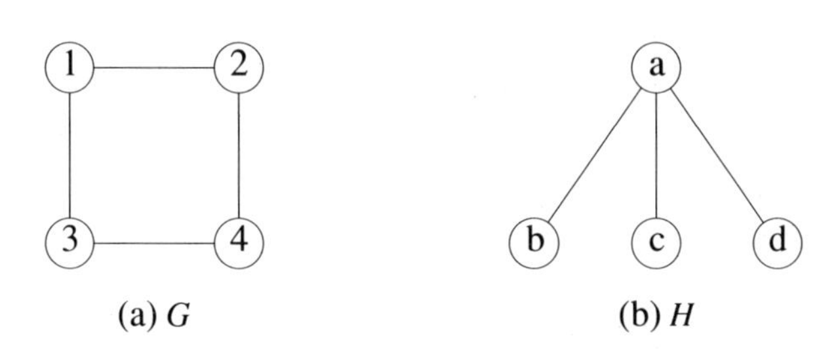

Figure 1.7 Example graphs

The ratio $|V(H)|/|V(G)|$ is called the *expansion*. The length of the longest path that any edge of G is mapped to is called the *dilation*. The *congestion* of any edge of H is the number of paths in the edge to path mapping that it is on. The maximum of the congestions of all edges of H is the *congestion* of the embedding. For the above example, the expansion is 1 while the dilation and congestion are both 2. In the remainder of this section the graph H will always be the graph corresponding to a hypercube interconnection. H_d will denote the hypercube graph with dimension d. This graph has 2^d vertices and each vertex has degree d.

1.2.2 Chains and Rings

Let G be a ring with 2^d vertices. Assume that the ring vertices are numbered 0 through 2^d-1 in ring order. Figure 1.3 (a) shows an eight processor ring. G can be embedded in H_d using a Gray code scheme (Chan and Saad 1986, and Johnsson 1987). The Gray code S_k is defined recursively as below:

$$S_1 = 0, 1; S_{k+1} = 0[S_k], 1[S_k]^R, k > 1$$

where $0[S_k]$ is obtained by prefixing each entry in S_k with a 0 and $1[S_k]^R$ is obtained by first reversing the order of the entries in S_k and then prefixing each entry with a 1. So, S_2 = 00, 01, 11, 10 and S_3 = 000, 001,

011, 010, 110, 111, 101, 100.

It is easy to verify that any two adjacent entries in S_k differ in exactly one bit. Furthermore, the first and last entries also differ in exactly one bit. Let $gray(i,k)$ be the i'th entry in S_k, $0 \le i < 2^k$. We shall use the following embedding of G into H_d:

(1) Vertex i of G is mapped to vertex $gray(i,d)$ of H, $0 \le i < 2^d$.

(2) Each edge (i, j) of G is mapped to the unique edge in H_d that connects vertices $gray(i,d)$ and $gray(j,d)$.

The expansion, dilation, and congestion factors for this embedding are all 1. Note that the same embedding may be used when G is a chain rather than a ring. It should also be noted that when the hypercube vertices are considered in Gray code order, every block of 2^j vertices forms a ring, $j \ge 0$ (blocks are obtained by partitioning the vertices so ordered into groups of size 2^j). The above Gray code embedding is particularly useful in the case of MIMD hypercubes as in such hypercubes each processor can send data to its neighbor on the chain in a single unit route. Such a data transfer takes d unit routes on an SIMD hypercube as data transfer along all dimensions of the hypercube are required.

1.2.3 Meshes

The ring embedding may be generalized to obtain an embedding of a $P \times Q$ mesh when P and Q are both powers of 2. Let $P = 2^p$ and $Q = 2^q$. We shall embed the mesh into H_d where $d = p + q$. The vertex in position (i, j) of the mesh is mapped to the hypercube vertex with binary representation $gray(i, p)gray(j, q)$. Figure 1.3 (b) shows the resulting mapping for a 2×4 mesh. In this case $p = 1$, and $q = 2$. So, H_3 is used and each processor index has three bits. The first bit in the vertex to which (i, j) is mapped is $gray(i, 1)$ and the last two are $gray(j, 2)$. It is easy to see that for each edge $((i, j), (k, l))$ in the mesh there is a unique edge in H_d that connects vertices $gray(i, p)gray(j, d)$ and $gray(k, p)gray(l, q)$ in H_d. This is true even if the mesh has row and column wrap around connections (i.e., the end of a row or column connects to its front).

The above mapping has expansion, dilation, and congestion of 1. An arbitrary $P \times Q$ mesh (i.e., one in which P and Q are not necessarily powers of 2) can be embedded in a hypercube with expansion 2 and dilation 2 (Chan 1986).

1.2.4 Full Binary Trees

Let T_i be the full binary tree of height i. Since T_i has 2^i-1 vertices, the best we can do is embed T_i into H_i, $i > 1$. Note that T_1 is trivially embedded into H_0. While T_2 can be embedded into H_2 with a dilation and congestion of 1 (Figure 1.8), there is no dilation 1 embedding of T_i into H_i for $i > 2$. However, T_i can be embedded into H_{i+1}, $i > 2$ with dilation 1 and into H_i, $i > 2$ with dilation 2.

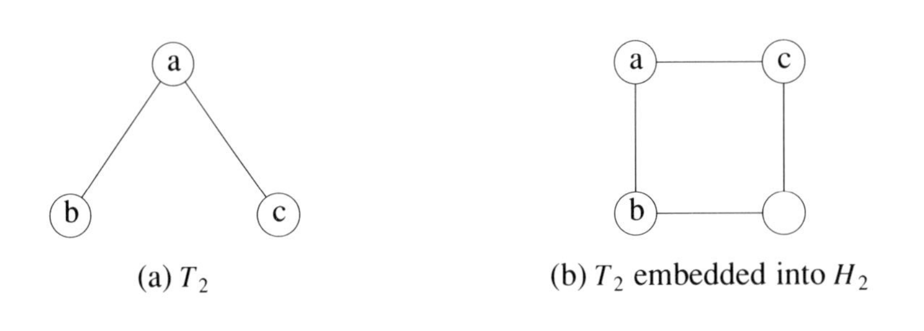

Figure 1.8 Dilation 1 embedding of T_2 into H_2

Theorem 1.1 (Wu 1985) There is no dilation 1 embedding of T_i into H_i for $i > 2$.

Proof: Suppose there is a dilation one embedding of T_i into H_i for some i, $i > 2$. Since H_i is symmetric, we may assume that the root of T_i (assumed to be at level 1) is mapped to vertex 0 of H_i. Its children (i.e., level two vertices of T_i) must be mapped to hypercube vertices that have exactly one 1 in their binary representation (as this is a dilation one embedding). The level three vertices of T_i must therefore be mapped to hypercube vertices with either zero or two ones. Hence the level four vertices must be mapped to hypercube vertices with one or three ones; and so on. So, the dilation one embedding must satisfy the following:

(1) Vertices of T_i that are on an odd level are mapped to hypercube vertices that have an even number of ones.

(2) Vertices of T_i that are on an even level are mapped to hypercube vertices that have an odd number of ones.

The number of hypercube vertices with an odd number of ones exactly equals the number with an even number of ones. So, H_i has 2^{i-1} vertices in each category. If i is even, then T_i has $2 + 2^3 + 2^5 + \cdots + 2^{i-1} = 2(2^i - 1)/3 > 2^{i-1}$ (for $i > 2$) vertices on even levels. Hence, H_i doesn't have enough vertices with an odd number of ones to accomodate the vertices of T_i that are on even levels.

If i is odd, then T_i has $2^0 + 2^2 + 2^4 + \cdots + 2^{i-1} = (2^{i+1} - 1)/3 > 2^{i-1}$ (for $i > 1$) vertices on odd levels. So, H_i doesn't have enough vertices with an even number of ones to accomodate the vertices of T_i that are on odd levels.

So, there can be no dilation one embedding of T_i into H_i for $i > 2$. □

Theorem 1.2 (Wu 1985) There is a dilation 1 embedding of T_i into H_{i+1} for $i > 0$.

Proof: The proof is by construction. The embeddings we shall construct will satisfy the following *free neighbor property*:

> *In the embedding of T_i into H_{i+1} the root of T_i is mapped to a vertex in H_{i+1} with the property that at least one of its hypercube neighbors, A, is free (i.e., no tree vertex is mapped to it) and furthermore A has a free neighbor B.*

Dilation 1 free neighbor embeddings of T_1 and T_2 into H_2 and H_3, respectively, are given in Figure 1.9.

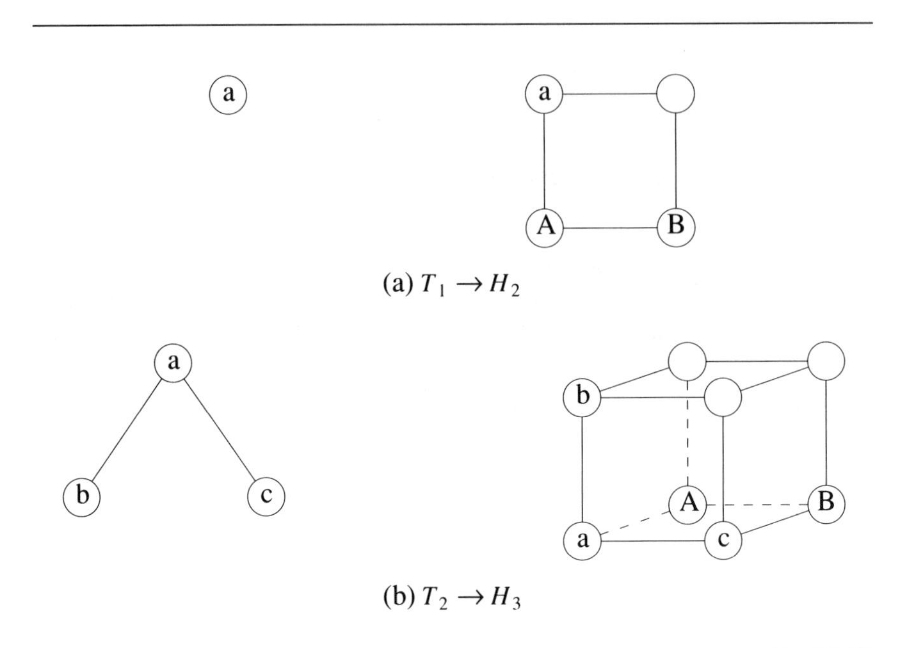

Figure 1.9 Free neighbor embeddings

From a dilation one free neighbor embedding of T_{d-1} into H_d we can obtain a dilation one free neighbor embedding of T_d into H_{d+1} in the following way:

(1) Let $0H_d$ denote the dimension d hypercube in H_{d+1} comprised solely of vertices whose most significant bit is 0. Let $1H_d$ denote the dimension d hypercube formed by the remaining vertices of H_{d+1}.

(2) Use the dilation one free neighbor mapping of T_{d-1} into H_d to map the left subtree of T_d into $0H_d$ and the right subtree of T_d into $1H_d$. Let 0A, 0B, 1A, and 1B denote the corresponding free neighbors.

(3) Apply a rigid transformation (i.e., a rotation and/or translation that does not change the relative positions of vertices in the mapping) to the mapping of the right subtree of T_d into $1H_d$ such that the root of this subtree becomes a neighbor of 0A and the transformed position $T(1\text{A})$ is a neighbor of 0B (see Figure 1.10 for the case $d = 3$).

(4) The root of T_d is mapped to 0A.

One may verify that this mapping has dilation one. Since 0A has the free neighbor 0B which in turn has the free neighbor $T(1A)$, the mapping satisfies the free neighbor property. □

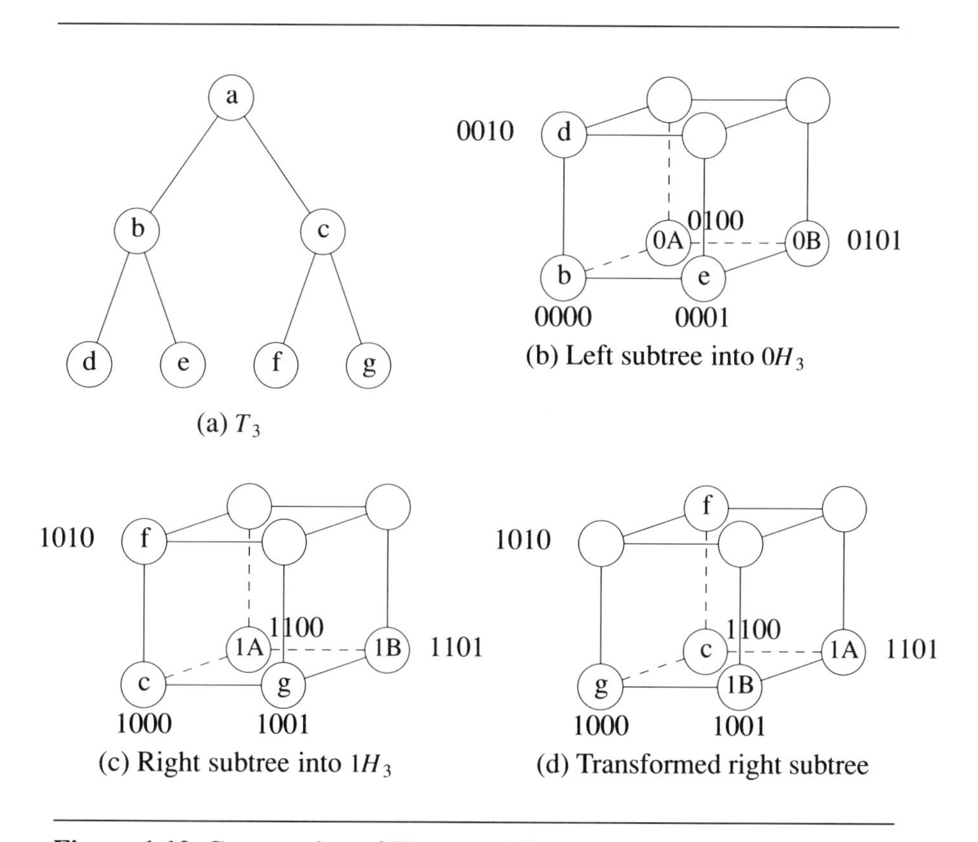

Figure 1.10 Construction of Theorem 1.2

Theorem 1.3 (Wu 1985) There is a dilation 2 embedding of T_i into H_i for $i > 0$.

Proof: The embedding of T_i into H_i will satisfy the following two properties:

(1) *Cost 2 Property*: Let A be the root of T_i and let L and R, respectively, be the roots of its left and right subtrees. The distance between the vertices that A and L are mapped to in H_i is 2 while that between the vertices that A and R are mapped to is 1.

(2) *Free Neighbor Property*: The sole free node in H_i is a neighbor of the node to which the root of T_i is mapped.

Figure 1.11 shows the embedding of T_1 and T_2 into H_1 and H_2, respectively. These embeddings satisfy both the above properties.

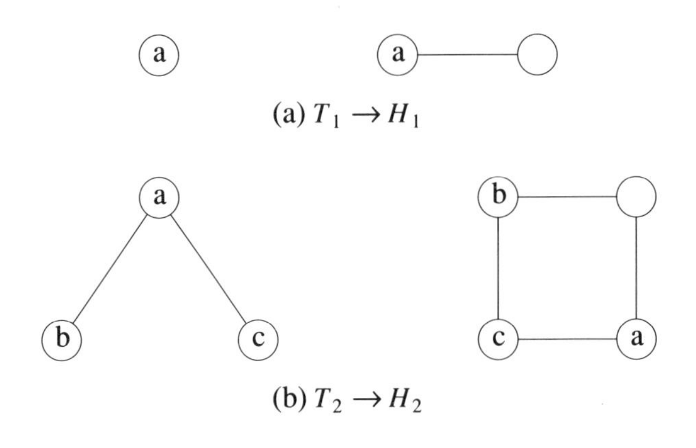

Figure 1.11 Dilation 2 embeddings for Theorem 1.3

From a dilation two embedding of T_{d-1} into H_{d-1} that satisfies the above two properties we can obtain a dilation two embedding of T_d into H_d that also satisfies these properties. The steps are:

(1) Embed the left subtree of T_d into $0H_{d-1}$. Let 0A be the vertex in H_d to which the root of the left subtree is mapped and let 0B be its free neighbor.

(2) Embed the right subtree of T_d into $1H_{d-1}$. Let 1A be the vertex in H_d to which the root of the left subtree is mapped and let 1B be its free neighbor.

(3) Map the root of T_d to the vertex 1B.

One may verify that this embedding satisfies the stated properties.

□

1.3 Performance Measures

The performance of uniprocessor algorithms and programs is typically measured by their time and space requirements. These measures are also used for multicomputer algorithms. Other measures that we shall now define are also used. Let t_p and s_p, respectively, denote the time and space required on a p node multicomputer. While s_p will normally be the total amount of memory required by a p node multicomputer, for distributed memory multicomputers (as is our hypercube) it is often more meaningful to measure the maximum local memory requirement of any node. This is because most multicomputers have equal size local memory on each processor.

To determine the effectiveness with which the multicomputer nodes are being used, one also measures the quantities *speedup* and *efficiency*. Let t_0 be the time required to solve the given problem on a single node using the conventional uniprocessor algorithm. Then, the *speedup*, S_p, using p processors is:

$$S_p = \frac{t_0}{t_p}$$

Note that t_1 may be different from t_0 as in arriving at our parallel algorithm, we may not start with the conventional uniprocessor algorithm.

The *efficiency*, E_p, with which the processors are utilized is:

$$E_p = \frac{S_p}{p}$$

Barring any anomalous behavior as reported in Kumar, Nageshwara, and Ramesh (1988), Lai and Sahni (1984), Li and Wah (1986), and Quinn and Deo (1986), the speedup will be between 0 and p and the efficiency between 0 and 1. To understand the source of anomalous behavior that results in $S_p > p$ and $E_p > 1$, consider the search tree of Figure 1.12. The problem is to search for a node with the characteristics of C. The best uniprocessor algorithm (i.e., the one that works best on most instances) might explore subtree B before examining C. A two processor parallelization might explore subtrees B and C in parallel. In this case, $t_2 = 2$ (examine A and C) while $t_0 = k$ where $k-1$ is the number of nodes in subtree B. So, $S_2 = k/2$ and $E_2 = k/4$.

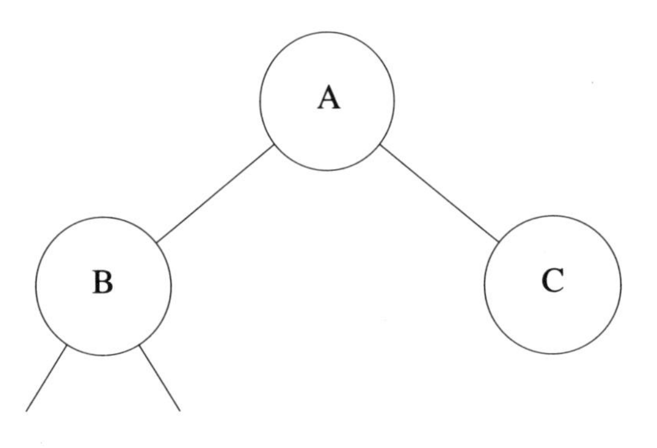

Figure 1.12 Example search tree

One may argue that in this case t_0 is really not the smallest uniprocessor time. We can do better by a breadth first search of the tree. In this case, $t_0 = 3$, $t_2 = 2$, $S_2 = 1.5$, and $E_2 = 0.75$. Unfortunately, given a search tree there is no known method to predict the optimal uniprocessor search strategy. Thus in the example of Figure 1.12, we could instead be looking for a node D that is at the bottom of the leftmost path from the root A. So, it is customary to use for t_0 the run time of the algorithm one would normally use to solve that problem on a uniprocessor.

While measured speedup and efficiency are useful quantities, neither give us any information on the scalability of our parallel algorithm to the case when the number of processors/nodes is increased from that currently available. It is clear that, for any fixed problem size, efficiency will decline as the number of nodes increases beyond a certain threshold. This is due to the unavailability of enough work, i.e., processor starvation. In order to use increasing numbers of processors efficiently, it is necessary for the work load (i.e., t_0) and hence problem size to increase also (Gustafson 1988). An interesting property of a parallel algorithm is the amount by which the work load or problem size must increase as the number of processors increases in order to maintain a certain efficiency or speedup. Kumar, Nageshwara, and Ramesh (1988) have introduced the concept of isoefficiency to measure this property. The *isoefficiency*, $ie(p)$, of a parallel algorithm/program is the amount by which the work load must increase to maintain a certain efficiency.

We illustrate these terms using matrix multiplication as an example. Suppose that two $n \times n$ matrices are to be multiplied. The problem size is n. Assume that the conventional way to perform this product is by using the classical matrix multiplication algorithm of complexity $O(n^3)$. Then, $t_0 = cn^3$ and the work load is cn^3. Assume further that p divides n. Since the work load can be evenly distributed over the p processors when $p \leq n^2$,

$$t_p = \frac{t_0}{p} + t_{com}$$

where t_{com} represents the time spent in interprocessor communication. So, $S_p = t_0/t_p = pt_0/(t_0+pt_{com})$ and $E_p = S_p/p = t_0/(t_0+pt_{com}) = 1/(1+pt_{com}/t_0)$. In order for E_p to be a constant, pt_{com}/t_0 must be equal to some constant $1/\alpha$. So, $t_0 =$ work load $= cn^3 = \alpha pt_{com}$. In other words, the work load must increase at least at the rate αpt_{com} to prevent a decline in efficiency. If t_{com} is ap (a is a constant), then the work load must increase at a quadratic rate. To get a quadratic increase in the work load, the problem size n needs increase only at the rate $p^{2/3}$ (or more accurately, $(a\alpha/c)^{1/3}p^{2/3}$).

Barring any anomalous behavior, the work load t_0 for an arbitrary problem must increase at least linearly in p as otherwise processor starvation will occur for large p and efficiency will decline. Hence, in the absence of anomalous behavior, $ie(p)$ is $\Omega(p)$. Parallel algorithms with smaller $ie(p)$ are more scalable than those with larger $ie(p)$.

The concept of isoefficiency is useful because it allows one to test parallel programs using a small number of processors and then predict the performance for a larger number of processors. Thus it is possible to develop parallel programs on small hypercubes and also do a performance evaluation using smaller problem instances than the production instances to be solved when the program is released for commercial use. From this performance analysis and the isoefficiency analysis one can obtain a reasonably good estimate of the program's performance in the target commercial environment where the multicomputer may have many more processors and the problem instances may be much larger. So with this technique we can eliminate (or at least predict) the often reported observation that while a particular parallel program performed well on a small multicomputer it was found to perform poorly when ported to a large multicomputer.

Chapter 2

Fundamental Operations

2.1 Data Broadcasting

In a data broadcast we begin with data in the A register of processor zero of the hypercube. This data is to be transmitted to the A register of each of the remaining processors in the hypercube. This may be accomplished using the binary tree transmitting scheme of Figure 2.1. This figure is for the case of a dimension 3 hypercube. The data to be broadcast is initially only in processor 0 (root of the broadcast tree). It is transmitted along bit 0 to processor 1 (001). This is denoted by the arrow from the root to its right child. Now processors 0 and 1 (level 2 processors) have the data. Both transmit along bit 1. The data is now contained in all level 3 processors (0, 2, 1, 3). These processors now send the data to their neighbor processors along bit 2. As a result, all processors in the hypercube have the data.

The code of Program 2.1 uses the broadcast tree idea to perform a data broadcast in a hypercube of dimension d. This code requires half the hypercube processors to send data in each transmittal step. While this is actually required only in the last transmittal step, the code is slightly simplified by permitting half the processors to transmit in each step.

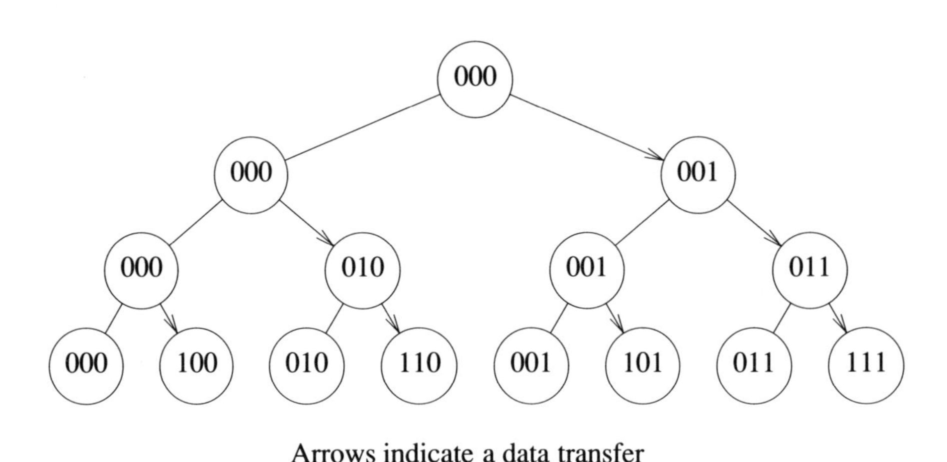

Arrows indicate a data transfer

Figure 2.1 Broadcasting in a dimension 3 hypercube

procedure *Broadcast* (A, d);
{Broadcast data in the A register of PE 0 to the remaining}
{PE's of the dimension d hypercube.}
  for $i := 0$ **to** $d - 1$ **do**
    $A(j^{(i)}) \leftarrow A(j),\ (j_i = 0)$;
end; {of *Broadcast*}

Program 2.1 Data broadcast

Note that if the data to be broadcast originates in processor k then the broadcast can be accomplished by modifying the selectivity function of Program 2.1 from $(j_i = 0)$ to $(j_i = k_i)$. In case we have available a special value *null* that is not otherwise an allowable value for A, then an alternate (and simpler) broadcasting scheme is possible. In this we begin by setting A to *null* in all processors other than the one that contains the data to be broadcast. The selectivity function of Program 2.1 is changed to $(A(j) \neq null)$. The resulting broadcasting algorithm is given in Program 2.2. The complexity of all versions of the broadcasting algorithm that we have discussed is O(d).

---

procedure *Broadcast* (A, d, k);
{Broadcast data in the A register of PE k to the remaining}
{PE's of the dimension d hypercube.}
{Assume there is a special value *null*}
  $A(j) := null, \ (j \neq k)$;
  for $i := 0$ **to** $d - 1$ **do**
    $A(j^{(i)}) \leftarrow A(j), \ (A(j) \neq null)$;
end; {of *Broadcast*}

Program 2.2 Data broadcast using a *null* value

2.2 Window Broadcast

Suppose that a dimension d hypercube is partitioned into windows of size 2^k processors each. Assume that this is done in such a way that the processor indices in each window differ only in their least significant k bits. As a result, the processors in each window form a subhypercube of dimension k. In a window broadcast we begin with data in the A register of a single processor in each window. This is the same relative processor for each window (i.e., its least significant k bits are the same across all windows). In each window the data from the window's originating processor is to be broadcast to all remaining processors in the window.

As an example consider the case $d = 3$, $k = 2$, and the originating PE in each window has bits 0 and 1 equal to 0. We have two windows of 4 processors each. The processors in one window have their most significant bit equal to 0 while those in the other window have this bit equal to one. The originating processor for the first window is 0 while that for the second is 4. Data from processor 0 is to be broadcast to processors 1, 2, and 3 and data from processor 4 is to be broadcast to processors 5, 6, and 7.

A window broadcast can be done in O(k) time using Program 2.3. This is almost identical to Program 2.1.

---

procedure *WindowBroadcast* (A, k);
{Broadcast in windows of 2^k processors. The originating processor in}
{each window has least significant bits $m_0, m_1, \cdots, m_{k-1}$.}
{Each window is a subhypercube.}
  for $i := 0$ **to** $k-1$ **do**
    $A(j^{(i)}) \leftarrow A(j), \; (j_i = m_i)$;
end; {of *WindowBroadcast*}

Program 2.3 Window broadcast

As in the case of Program 2.1, Program 2.3 can be generalized to broadcast data from different relative processors in different windows by using a special value *null*. This generalization is given in Program 2.4. The procedure assumes that initially *selected* (j) is true iff processor j is the originating processor for its window. The complexity of Program 2.4 is also O(k).

2.3 Data Sum

Assume that a dimension d hypercube is partitioned into subhypercubes (or windows) of dimension k as in the case of window broadcast. The data sum operation sums the A register data of all processors in the same window. The result is left in the A register of the processor with least index in each window. I.e., data sum computes:

$$sum(iW) = \sum_{j=0}^{W-1} A(iW+j), \quad 0 \le i < P/W$$

where $W = 2^k$ and $P = 2^d$.

```
procedure WindowBroadcast (A, k);
{Broadcast in windows of 2^k processors. The originating processor in}
{each window has selected = true. The special value null is used.}
{Each window is a subhypercube.}
  A (j) := null, (selected (j) = false);
  for i := 0 to k - 1 do
    A (j^(i)) <- A (j), (A (j) ≠ null);
end; {of WindowBroadcast}
```

Program 2.4 Window broadcast from arbitrary processors using *null* value

The data sum operation can be performed in O(k) time by first summing in subwindows of size 2, then of size 4, and so on until the subwindow size becomes W. Consider the example of Figure 2.2. This shows the computation of data sum on an eight processor SIMD hypercube. The window size is also eight. The leaf nodes represent the initial configuration. The number outside a node is the index of the processor it represents and the number inside a node is the A value of the corresponding processor. First, pairs of processors add their A values. For this the odd processor in each pair transmits its A value to the even processor in the pair. This data transfer is shown by arrows in Figure 2.2. Following the addition in each pair we have the configuration of the next level of Figure 2.2. In the next step the processor pairs (0, 2) and (4, 6) compute their sums. This corresponds to the sum of the initial data in the two subwindows of size 4. Processor 2 transmits the sum it computed in the previous step to processor 0 and processor 6 tranmits this sum to processor 4. Now processor 4 transmits its subwindow sum (16) to processor 0 which computes the overall sum 26.

The code is given in Program 2.5. Each processor has a variable *awake* which tells it whether or not it is still involved in the summing process. Initially all processors are awake. After the first data transfer step only the even processors remain awake. Following the next data transfer

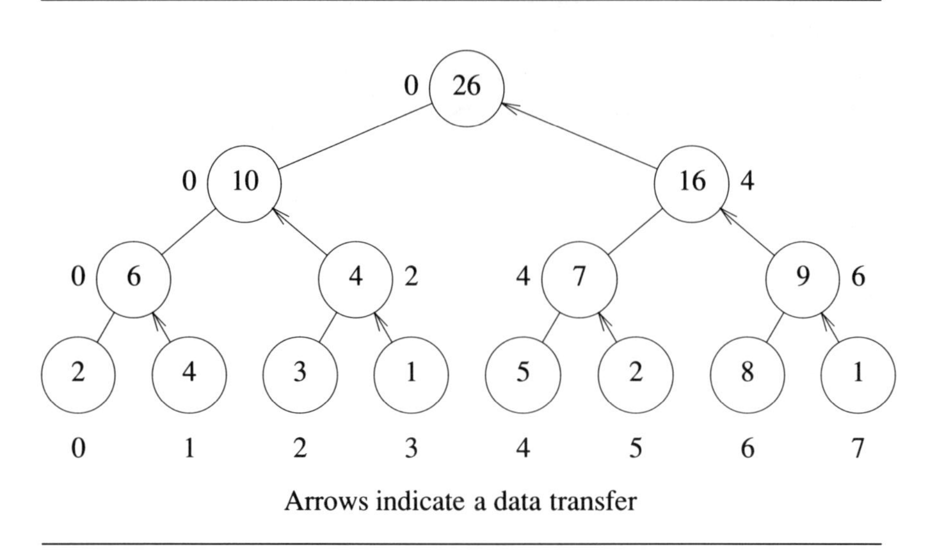

Figure 2.2 Data sum in a dimension 3 hypercube

step only processors with their two least significant bits equal to zero remain awake and so on. In the end only processor 0 is awake. The complexity of *SIMDDataSum* is $O(k)$.

By removing the selection functions from Program 2.5 we get a procedure which computes the window sum in each processor of the hypercube. This is given in Program 2.6. It is actually slightly easier to have all processors compute the window sum than to just have a single processor do this as in Program 2.5.

```
procedure SIMDDataSum (A, k);
{Sum the A register data in windows of 2^k processors.}
{The sum is left in the window processor with least index.}
{Each window is a subhypercube.}
   awake (j) := true;
   for i := 0 to k - 1 do
   begin
      B (j^(i)) ← A (j), ((j_i = 1) and awake (j));
      awake (j) := false, (j_i = 1);
      A (j) := A (j) + B (j), (awake (j));
   end;
end; {of SIMDDataSum}
```

Program 2.5 Data sum in subhypercubes of dimension k

```
procedure SIMDAllSum (A, k);
{Sum the A register data in all processors of windows of 2^k processors.}
{Each window is a subhypercube.}
   for i := 0 to k - 1 do
   begin
      B (j^(i)) ← A (j);
      A (j) := A (j) + B (j);
   end;
end; {of SIMDAllSum}
```

Program 2.6 Procedure to have all window processors compute the sum

2.4 Prefix Sum

The assumptions and initial condition are the same as for the data sum operation. If $l = iW + q$, $0 \le q < W$, is a processor index, then processor l is the q'th processor in window i. This processor is to compute:

$$S(iW + q) = \sum_{j=0}^{q} A(iW + j), \quad 0 \le i < P/W, 0 \le q < W$$

The prefix sums in windows of size $W = 2^k$ may be easily computed if we know the following values in each of the size 2^{k-1} subwindows that make up the 2^k window:

(1) Prefix sums in the 2^{k-1} subwindow

(2) Sum of all A values in the 2^{k-1} subwindow

The prefix sums relative to the whole size W window are obtained as below:

(1) If a processor is in the left 2^{k-1} subwindow, then its prefix sum is unchanged.

(2) The prefix sum of a processor in the right subwindow is its prefix sum when considered as a member of a 2^{k-1} window plus the sum of the A values in the left subwindow.

Figure 2.3 gives an example prefix computation that uses this strategy. The number of processors and the window size $W = 2^k$ are both 8. Line 0 gives the initial A values. The prefix sums in the current windows are stored in the S registers and the sum of the A values of the processors in the current windows are stored in the T registers. We begin with windows of size 1. The initial S and T values are given in lines 1 and 2, respectively. Next, the S and T values for windows of size 2 are obtained. These are given in lines 3 and 4. Line 5 and 6 give the S and T values when the window size is 4 and lines 9 and 10 give these values for the case when the window size is 8. Program 2.7 is the resulting procedure. Its complexity is $O(k)$.

	PE								
line	0	1	2	3	4	5	6	7	
0	2	4	3	1	5	2	8	1	A
1	2	4	3	1	5	2	8	1	S
2	2	4	3	1	5	2	8	1	T
3	2	6	3	4	5	7	8	9	S
4	6	6	4	4	7	7	9	9	T
5	2	6	9	10	5	7	15	16	S
6	10	10	10	10	16	16	16	16	T
7	2	6	9	10	15	17	25	26	S
8	26	26	26	26	26	26	26	26	T

Figure 2.3 Example to compute prefix sums in an SIMD hypercube

2.5 Shift

The operation *shift* (A, i, W) shifts the data in the A register of the processors counterclockwise by i processors in windows of size W where W is a power of 2. In the case of an SIMD hypercube of size $P = 2^d$ the processors in counterclockwise order are 0, 1, 2, $\cdots$, $P-1$, 0. So a shift of i with a window size of $W = P$ results in moving data from PE(j) to PE($(j+i)$ **mod** P). In the case of an MIMD hypercube the PE ordering corresponds to that obtained by mapping a chain into a hypercube using a Gray code (Section 1.2.2). The counterclockwise ordering is $gray(0,d)$, $gray(1,d)$, $\cdots$, $gray(P-1,d)$, $gray(0,d)$.

---

procedure *SIMDPrefixSum* (A, k, S);
{Compute the prefix sums of A in windows of size 2^k}
begin
 {Initialize for size 1 windows}
 $S(i) := A(i)$;
 $T(i) := A(i)$;

 {Compute for size 2^{b+1} windows}
 for $b := 0$ **to** $k-1$ **do**
 begin
  $B(i^{(b)}) \leftarrow T(i)$;
  $S(i) := S(i) + B(i),\ (i_b = 1)$;
  $T(i) := T(i) + B(i)$;
 end;
end; {of *SIMDPrefixSum*}

Program 2.7 SIMD prefix sums

2.5.1 SIMD Shift

The strategy (Kumar and Krishnan 1987, Ranka and Sahni 1988b) is to reduce a shift in a window of size W into two independent shifts in windows of size $W/2$. By using this repeatedly we end up with shifts in windows of size one. Shifts in windows of this size are equivalent to null shifts. Let the current window size be $2M$ (W is the initial window size and $M = W/2$ at the start). For the reduction, we consider two cases: (1) $0 < i \le M$ and (2) $M < i < 2M$. In case i is not in this range, it may be replaced by i **mod** $2M$.

(1) First consider an example. Suppose that $M = 4$ and $i = 3$. Then the initial configuration of line 1 of Figure 2.4 is to be transformed into that of line 2. The 8 processors may be partitioned into two windows of size 4 each. The left window consists of processors 0 through 3 and processors 4 through 7 make up the right window. Examining the initial and final configurations, we notice that b, c,

and d are initially in the left window and they are in the right window after the shift. Also, f, g, and h are initially in the right window and are in the left window following the shift. If we exchange b, c, and d with f, g, and h, respectively, then we obtain the configuration of line 3. Now each of the two windows of size 4 has the data it needs for its final configuration. Furthermore, a shift of 3 in each window will result in the final configuration.

	PE							
line	0	1	2	3	4	5	6	7
1	a	b	c	d	e	f	g	h
2	f	g	h	a	b	c	d	e
3	a	f	g	h	e	b	c	d
4	c	d	e	f	g	h	a	b
5	e	f	c	d	a	b	g	h

Figure 2.4 SIMD shift example

To perform a shift of i, $0 < i \leq M$, in a window of size $W = 2M$, we consider its two subwindows (left and right) of size M each. Following the shift some of the elements initially in the left subwindow are found in the right subwindow and an equal number initially in the right subwindow are found in the left subwindow. Let A and C, respectively, denote the elements of the left and right subwindows that do not change subwindows as a result of the shift. Let B

and D, respectively, denote the remaining elements of the left and right subwindows (see Figure 2.5). One may verify that B and D each contain i elements and both consist of elements beginning at position $first = M - i$ and ending at position $last = M - 1$ relative to the start of the respective subwindow. It is easy to see that if we exchange B and D, then the desired final configuration can be obtained by performing a shift of i **mod** M in each of the size M subwindows. The **mod** M is necessary to handle the case $i = M$.

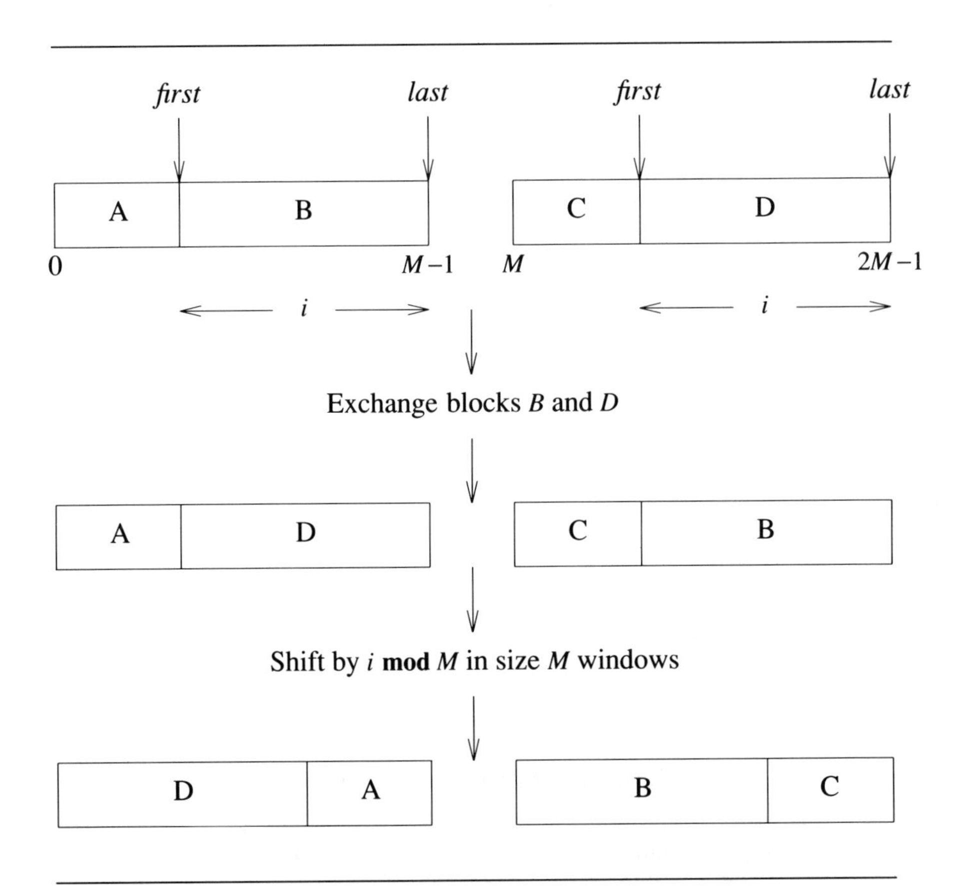

Figure 2.5 Steps in performing a shift of i, $0 < i \leq M$

(2) Consider the initial configuration of line 1 of Figure 2.4. This time assume that $i = 6$. The final configuration is shown in line 4. Now a, b, e, and f change windows between the initial and final configurations.

As before, we consider the elements in the two subwindows of the size W window. Let B and D, respectively, denote the elements of the left and right subwindows that do not change subwindows as a result of the shift. Let A and C, respectively, denote the remaining elements of the left and right subwindows (see Figure 2.6). One may verify that A and C each contain $2M - i$ elements and both consist of elements beginning at position $first = 0$ and ending at position $last = 2M - i - 1$ of the respective subwindow. We see that if we exchange A and C, then the desired final configuration can be obtained by performing a shift of i **mod** M in each of the size M subwindows.

The preceding discussion results in the procedure of Program 2.8. Here *RightmostOne* (i) is a function that returns the least significant bit of i that is one. In case $i = 0$, it returns -1. So, *RightmostOne* $(12) = 2$ and *RightmostOne* $(9) = 0$. The complexity of *SIMDShift* is $O(\log W)$. In case the shift amount i is a power of 2, the complexity becomes $O(\log(W/i))$.

2.5.2 MIMD Shift

On an MIMD hypercube a shift of i is obtained by making a sequence of shifts each of which is a power of 2 shift (Ranka and Sahni 1988b). So, for example, a shift of 11 is obtained by performing shifts of 8, 2, and 1 in any order. A power of 2 shift can be made in two steps. Suppose we are to perform *SHIFT* (A,i,W) where i and W are both powers of 2. We may assume that $i < W$ (if not, replace i by i **mod** W). Each size W window is comprised of some number of size i subwindows. Because of the Gray code indexing, the processor indices in any subwindow differ from those in the adjacent subwindow in exactly one bit. Furthermore, after changing this bit, the indices in one subwindow are in the reverse

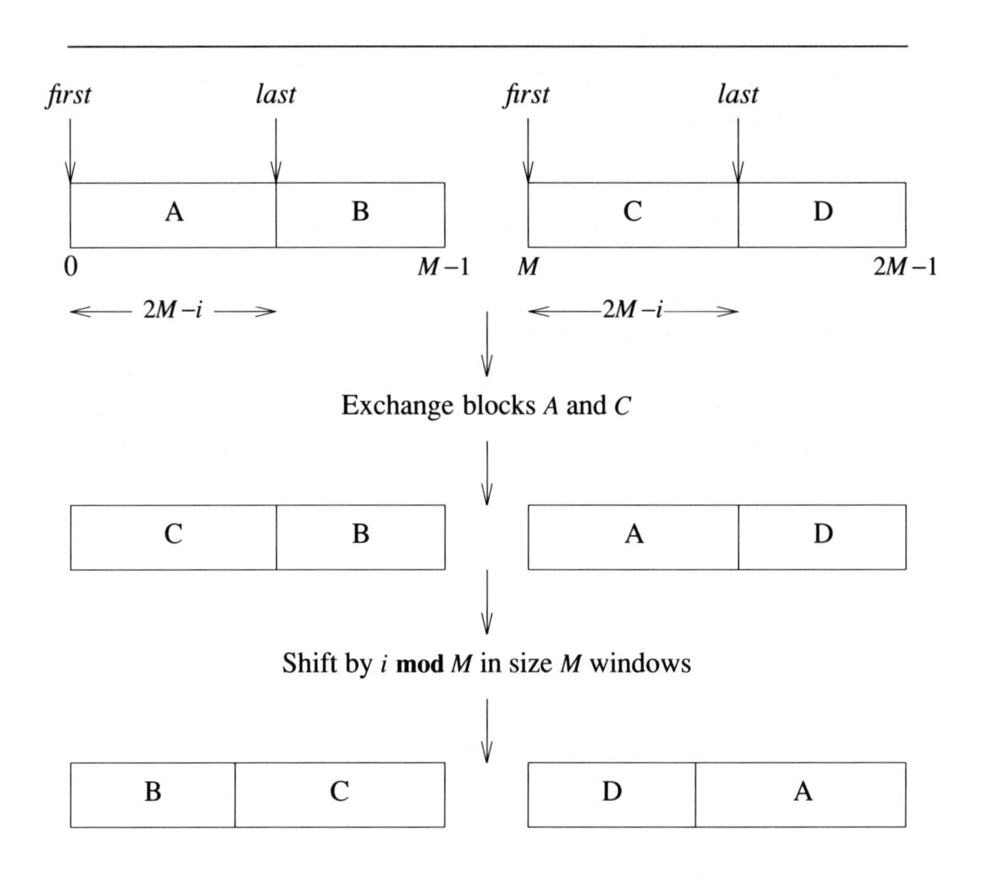

Figure 2.6 Steps in performing a shift of i, $M < i < 2M$

order of those in the adjacent window. In Figure 2.7 the processor indices of an eight processor hypercube are provided in Gray code order. The four windows of size two are defined by the processor pairs {000, 001}, {011, 010}, {110, 111}, and {101, 100}. Changing bit 1 of the first pair gives the pair {010, 011} which is the reverse of the second pair; changing bit 2 of the second pair gives {111, 110} which is the reverse of the third pair; changing bit 1 of the third pair gives the reverse of the fourth pair; and changing bit 2 of the fourth pair gives the reverse of the first pair.

```
procedure SIMDShift (A, i, W);
{Counterclockwise shift of A by i in windows of size W.}
{SIMD version.}
  i := i mod W;
  b := RightmostOne (i);
  M := W; k := log2M;
  for j := k-1 downto b do
  begin
    M := M div 2;
    if i <= M then begin
                     first := M - i;
                     last := M - 1;
                   end
              else begin
                     first := 0;
                     last := 2*M - i - 1;
                   end;
    {p is this processor's index, a is position in size M window}
    a := p mod M;
    A(p^(j)) ← A(p), (first ≤ a ≤ last);
    i := i mod M; {remaining shift}
  end; {of for loop}
end; {of SIMDShift}
```

Program 2.8 SIMD shift

With the above observation we arrive at the following two step algorithm to perform a shift of i when i is a power of 2:

Step 1 Each processor in a subwindow of size i routes to the clockwise adjacent subwindow of size i in each window of size W.

Step 2 If $i > 1$, then in each subwindow of size i the data is reversed by exchanging between the subwindows of size $i/2$ that make up the size i subwindow.

	PE							
line	000	001	011	010	110	111	101	100
1	a	b	c	d	e	f	g	h
2	h	g	f	e	d	c	b	a
3	e	f	g	h	a	b	c	d
4	h	g	b	a	d	c	f	e
5	g	h	a	b	c	d	e	f

Figure 2.7 MIMD shift example

Consider a shift of $i = 4$ in an eight processor hypercube with $W = 8$. Line 1 of Figure 2.7 gives the initial configuration. There are two subwindows of size 4 ({000, 001, 011, 010} and {110, 111, 101, 100}). The first subwindow routes to the second along bit 2 and the second routes to the first along bit 2. The result is given in line 2 of Figure 2.7. Now the data is in the right size 4 subwindow but in the reverse order. To get the right order data is exchanged between processors pairs of subwindows of size 2. Thus the processors in the pair ({000, 001}, {011, 010}) as well as those in the pair ({110, 111}, {101, 100}) exchange along bit 1. The result is the configuration of line 3 of the figure.

As another example, consider the case of $i = 2$ and $W = 8$. The initial configuration is that of line 1 of Figure 2.7. First, each size 2 subwindow sends its data to the size 2 subwindow that is counterclockwise from it. The subwindow {000, 001} transmits to {011, 010} along bit 1; {011,

010} transmits to {110, 111} along bit 2; {110, 111} transmits to {101, 100} along bit 1; and {101, 100} transmits to {000, 001} along bit 2. The result is shown in line 4 of Figure 2.7. To complete the shift, pairs of windows of size 1 that make up a size 2 window exchange. Following this exchange we get the configuration of line 5.

From the above it follows that when i is a power of 2 the shift can be done in O(1) time. When i is not a power of 2, a shift of i takes $O(\log_2 W)$ time as i can have at most $\log_2 W$ ones in its binary representation. We note that if the number of ones in the binary representation of i exceeds $\log_2 W/2$ then it is faster to perform a clockwise shift of $W - i$. The algorithm for clockwise shifts is similar to that for counterclockwise shifts.

2.6 Data Circulation

Consider a $P = 2^p$ processor hypercube. We are required to circulate the data in the A register of these PEs so that this data visits each of the P processors exactly once. An optimal circulation algorithm for MIMD hypercubes results by repeatedly shifting by 1. An optimal circulation for SIMD hypercubes results from the use of the exchange sequence X_p (Dekel, Nassimi, and Sahni 1981) defined as

$$X_1 = 0, \quad X_q = X_{q-1}, q-1, X_{q-1} \ (q > 1)$$

So, $X_2 = 0, 1, 0$ and $X_3 = 0, 1, 0, 2, 0, 1, 0$. The exchange sequence X_q essentially treats a q dimensional hypercube as two $q-1$ dimensional hypercubes. Data circulation is done in each of these in parallel using X_{q-1}. Next an exchange is done along bit $q-1$. This causes the data in the two halves to be swapped. The swapped data is again circulated in the two half hypercubes using X_{q-1}. Let $f(q, i)$ be the i'th number (left to right) in the sequence X_q, $1 \le i < 2^q$. So, $f(2,1) = 0$, $f(2,2) = 1$, $f(2,3) = 0$, etc. The resulting SIMD data circulation algorithm is given in Program 2.9.

---

procedure *circulate* (A);
{Data circulation in a $P = 2^p$ processor SIMD hypercube}
begin
  for i := 1 **to** $P - 1$ **do**
    $A(j^{f(p,\,i)}) \leftarrow A(j)$;
end; {of *circulate*}

Program 2.9 Data circulation in an SIMD hypercube

An example of data circulation in an eight processor hypercube is given in Figure 2.8. Line 0 gives the initial contents of the A registers and line i gives the contents following iteration i of the **for** loop of Program 2.9, $1 \le i < P$. Line i also gives the value of $f(3,i)$.

The function f can be computed by the control processor in $O(P)$ time and saved in an array of size $P-1$ (actually it is convenient to compute f on the fly using a stack of height $\log P$). The following theorem allows each processor to compute the origin of the current A value.

Theorem 2.1 (Ranka and Sahni 1988b) Let $A[0], A[1], \ldots, A[2^p - 1]$ be the values in $A(0), A(1), \ldots, A(2^p - 1)$ initially. Let $index(j, i)$ be such that $A[index(j, i)]$ is in $A(j)$ following the i'th iteration of the for loop of Program 2.9. Initially, $index(j, 0) = j$. For every i, $i > 0$, $index(j, i) = index(j, i-1) \oplus 2^{f(p,\,i)}$ ($\oplus$ is the exclusive or operator).

Proof: We prove this using mathematical induction. The theorem is trivially true for $i = 1$. Assume the theorem is true for i, $l \ge i > 0$. Thus

$$index(j, l) = index(j, l-1) \oplus 2^{f(p,\,l)}$$

$$= index(j, l-2) \oplus 2^{f(p,\,l-1)} \oplus 2^{f(p,\,l)}$$

$$= j \oplus 2^{f(p,\,1)} \cdots \oplus 2^{f(p,\,l-1)} \oplus 2^{f(p,\,l)}$$

	PE								
line	0	1	2	3	4	5	6	7	$f(3,i)$
0	a	b	c	d	e	f	g	h	
1	b	a	d	c	f	e	h	g	0
2	d	c	b	a	h	g	f	e	1
3	c	d	a	b	g	h	e	f	0
4	g	h	e	f	c	d	a	b	2
5	h	g	f	e	d	c	b	a	0
6	f	e	h	g	b	a	d	c	1
7	e	f	g	h	a	b	c	d	0

Figure 2.8 SIMD circulate example

Similarly

$$index\,(j^{(f(p,\,l+1))},\,l) = j^{(f(p,\,l+1))} \oplus 2^{f(p,\,1)} \cdots \oplus 2^{f(p,\,l-1)} \oplus 2^{f(p,\,l)}$$

Now from procedure *circulate* (Program 2.9), we have

$$index\,(j,\,l+1) = index\,(j^{(f(p,\,l+1))},\,l)$$

$$= j^{(f(p,\,l+1))} \oplus 2^{f(p,\,1)} \cdots \oplus 2^{f(p,\,l-1)} \oplus 2^{f(p,\,l)}$$

$$= j \oplus 2^{f(p,\, l+1)} \oplus 2^{f(p,\, 1)} \ldots \oplus 2^{f(p,\, l-1)} \oplus 2^{f(p,\, l)}$$

$$= j \oplus 2^{f(p,\, 1)} \ldots \oplus 2^{f(p,\, l)} \oplus 2^{f(p,\, l-1)} \oplus 2^{f(p,\, l+1)}$$

(Since $\oplus$ is commutative and associative)

$$= index(j,\, l) \oplus 2^{f(p,\, l+1)} \quad \square$$

Let us apply the theorem to processor 0 of Figure 2.8. Initially, we have *index* (0,0) = 0. From the theorem it follows that *index* (0,1) = $0 \oplus 2^{f(3,1)} = 0 \oplus 1 = 1$; *index* (0,2) = $1 \oplus 2^{f(3,2)} = 1 \oplus 2 = 3$; *index* (0,3) = $3 \oplus 2^{f(3,3)} = 3 \oplus 1 = 2$; *index* (0,4) = $2 \oplus 2^{f(3,4)} = 2 \oplus 4 = 6$; *index* (0,5) = $6 \oplus 2^{f(3,5)} = 6 \oplus 1 = 7$; etc.

Some of our later algorithms will require the circulating data to return to the originating PEs. This can be accomplished by a final data exchange along the most significant bit. For convenience, we define $f(p, 2^p) = p - 1$.

2.7 Even, Odd, And All Shifts

In some applications it is desirable to shift the *A* register data by all possible even distances, or all possible odd distances, or by all possible distances. However, the order in which the shifts are performed is not specified. For example, in an eight processor hypercube the possible even shifts are 2, 4, and 6. We are free to perform these in any order. So, the first shift may be 6, the next 2, and the last 4. In an MIMD hypercube the three sets of shifts can be performed easily in time linear in the total number of shifts. First, all shifts are obtained in linear time by repeatedly shifting by 1. To get all odd shifts, we repeatedly shift by 1 and use only every other shift. The same strategy obtains all even shifts. Performing all shifts in the above three categories in linear time on an SIMD hypercube is more difficult as we can transfer data on one dimension at a time only.

2.7.1 All Even Shifts

In a P processor hypercube there are exactly $(P/2) - 1$ different even shifts (recall that $P = 2^k$ is a power of 2). These may be assumed to be in the range $[1, P-1]$. An even shift, i, that is not in this range is equivalent to the the even shift i **mod** P which is either 0 or in the above range. A shift of zero is equivalent to no shift.

A *shift distance sequence*, E_k, is a sequence $d_1 d_2 \cdots d_{2^{k-1}-1}$ of positive integers such that a clockwise shift of d_1, followed by one of d_2, followed by one of d_3, etc. covers all even length shifts in the range $[1, P-1] = [1, 2^k-1]$. Note that $E_0 = E_1 = null$ as there are no even length shifts in the range $[1, 2^k-1]$ when $k = 0$ or 1. When $k = 2$, there is one even shift in the range $[1, 2^k-1] = [1, 3]$. This is a shift of 2. So, $E_2 = 2$. This transforms the length $P = 2^2$ sequence *abcd* into the sequence *cdab*. In general, the choice $E_k = 2, 2, 2, \cdots$ will serve to obtain all even length shifts. From the complexity stand point this choice is poor as each shift of 2 requires $O(\log(P/2))$ unit routes. Better performance is obtained by defining

$$E_0 = E_1 = null, \ E_2 = 2$$

$$E_k = InterLeave\,(E_{k-1}, 2^{k-1}), \ k>2$$

where *InterLeave* is an operation that inserts a 2^{k-1} in front of E_{k-1}, at the end of E_{k-1}, and between every pair of adjacent distances in E_{k-1}. Thus,

$$E_3 = Interleave\,(E_2, 4) = 4\ 2\ 4$$

$$E_4 = Interleave\,(E_3, 8) = 8\ 4\ 8\ 2\ 8\ 4\ 8$$

When the shift sequence E_k is used, the effective shift following d_i is $(\sum_{j=1}^{i} d_j)\ mod\ 2^k$. Thus when E_3 is used on the sequence *abcdefgh*, we get

d	sequence	effective shift
4	*efghabcd*	4
2	*cdefghab*	6
4	*ghabcdef*	2 = 10 mod 8

In the shift sequence E_k, shifts of $P/2$ are done $P/4$ times (recall that $P = 2^k$); shifts of $P/4$ are done $P/8$ times; shifts of $P/8$ are done $P/4$ times; etc. Since shifts of $P/2$ cost less than shifts of $P/4$ which in turn cost less than shifts of $P/8$, etc., we expect the shift sequence E_k to cost significantly less than performing a sequence of length 2 shifts.

Theorem 2.2 (Ranka and Sahni 1988b) Let $E[k, i]$ be d_i in the sequence E_k, $k \geq 2$. Let $ESUM[k, i] = (\sum_{j=1}^{i} E[k, j]) \bmod 2^k$. Then $\{ESUM[k, i] \mid 1 \leq i \leq 2^{k-1} - 1\} = \{2, 4, 6, 8, ..., 2^k - 2\}$.

Proof: The proof is by induction on k. The theorem is clearly true for $k = 2$. Let it be true for $k = l \geq 2$. We prove that it will be true for $k = l + 1$. Hence by induction it will be true for all values of k, $k \geq 2$.

We prove the following statements which prove the required result.

(1) $ESUM[l + 1, i] < 2^{l+1}$.

(2) $ESUM[l + 1, i]$ is even.

(3) $ESUM[l + 1, i] \neq 0$

(4) $ESUM[l + 1, i] \neq ESUM[l + 1, k]$ if $i \neq k$

(1) and (2) follow directly from the definitions of $ESUM[l + 1, i]$ and E_{l+1}. We prove (3) and (4) by contradiction. Suppose $ESUM[l + 1, i] = 0$ for some value of i, say $a > 1$ (the case of $a = 1$ is obvious as $E[l + 1, 1] \neq 0$). Then,

$$(\sum_{j=1}^{a} E[l + 1, j]) \bmod 2^{l+1} = 0$$

$$(\equiv)(\sum_{j=1}^{\lfloor a/2 \rfloor} E[l, j] + \lceil a/2 \rceil 2^l) \bmod 2^{l+1} = 0$$

$$(=>)\left(\sum_{j=1}^{\lfloor a/2 \rfloor} E[l, j]\right) \bmod 2^l = 0$$

$$(\equiv) ESUM[l, \lfloor a/2 \rfloor] = 0$$

Contradiction (as $\lfloor a/2 \rfloor > 0$ and $ESUM[l, i] > 0$ for $i > 0$).

If $k = i \pm 1$ then (4) is true by the definition of $ESUM$. Suppose $ESUM[l+1, i] = ESUM[l+1, k]$, $i \neq k$, $k \neq i \pm 1$. So,

$$\left(\sum_{j=1}^{i} E[l+1, j]\right) \bmod 2^{l+1} = \left(\sum_{j=1}^{k} E[l+1, j]\right) \bmod 2^{l+1}$$

$$(\equiv)\left(\sum_{j=1}^{\lfloor i/2 \rfloor} E[l, j] + \lceil i/2 \rceil 2^l\right) \bmod 2^{l+1} = \left(\sum_{j=1}^{\lfloor k/2 \rfloor} E[l, j] + \lceil k/2 \rceil 2^l\right) \bmod 2^{l+1}$$

$$(=>)\left(\sum_{j=1}^{\lfloor i/2 \rfloor} E[l, j]\right) \bmod 2^l = \left(\sum_{j=1}^{\lfloor k/2 \rfloor} E[l, j]\right) \bmod 2^l$$

$$(\equiv) ESUM[l, \lfloor i/2 \rfloor] = ESUM[l, \lfloor k/2 \rfloor]$$

Contradiction (as $\lfloor i/2 \rfloor \neq \lfloor k/2 \rfloor$). □

Theorem 2.3 (Ranka and Sahni 1988b) The shift sequence E_k can be done in $2(2^k - k - 1)$ unit routes, $k \geq 2$.

Proof: Procedure *SIMDShift* (Program 2.8) performs a power of 2 shift 2^i in a window of size 2^k using $2\log(2^k/2^i) = 2(k-i)$ unit routes (assuming unidirectional links). Let cost(E_k) be the number of unit routes required by the sequence E_k. The 2^{k-2}, 2^{k-1}'s in E_k take 2 routes each. The cost of the remaining shifts in E_k is cost(E_{k-1}) + $2(2^{k-2}-1)$. The additive term $2(2^{k-2}-1)$ accounts for the fact that each of the remaining $2^{k-2}-1$ routes is done in a window of size 2^k rather than 2^{k-1} (as assumed for E_{k-1}). Hence,

$$cost(E_k) = \begin{cases} cost(E_{k-1}) + 2(2^{k-2} - 1) + 2 * 2^{k-2}, & k > 2 \\ 2, & k = 2 \end{cases}$$

So, $cost(E_k) = 2(2^k - k - 1)$, $k \geq 2$. □.

The result of the preceding theorem is important as it says that the average cost of rotation in E_k is $\frac{2(2^k - k - 1)}{2^{k-1} - 1} < 4$. So, we can perform even length rotations with O(1) average cost.

2.7.2 All Odd Shifts

All odd shifts can be performed in linear time by first performing a shift of 1 and then performing all even shifts. An alternate is to develop a shift sequence (like E_k) that obtains all odd shifts.

2.7.3 All Shifts

Let F_k be the sequence obtained by dividing each distance in E_k by 2. So, $F_0 = F_1 = null$, $F_2 = 1$, $F_3 = 2, 1, 2$, etc.

Theorem 2.4 (Ranka and Sahni 1988b) Let $F[k, j]$ be the j'th distance in F_k and let $FSUM[k, i] = (\sum_{j=1}^{i} F[k, j]) \bmod 2^{k-1}$.

(1) $\{FSUM[k, i] \mid 1 \leq i \leq 2^{k-1} - 1\} = \{1, 2, 3, ..., 2^{k-1} - 1\}$

(2) All the shifts in F_k can be done in a window of size 2^{k-1} in $2(2^k - k - 1)$ unit routes.

Proof: Similar to the proof of the preceding two theorems. □

2.8 Consecutive Sum

Suppose that a hypercube is tiled by windows of size M where M is a power of 2 and that each processor has the M values $X[0..M-1]$. The consecutive sum operation is performed in windows of size M. The j'th processor in such a window is to compute the sum of the $X[j]$ values in the M processors in its window. I.e., the j'th processor in a window computes:

$$S(j) = \sum_{i=0}^{M-1} X[j](i), \quad 0 \le j < M$$

where i refers to the i'th processor in the window.

The S's can be computed by having the j'th processor of the window originate an S token that is initialized to $X[j](j)$, $0 \le j < M$ (Ranka and Sahni 1988d). The S tokens are then circulated through the size M window. Each time the token that originated in the j'th processor of the window reaches a processor it adds to itself the $X[j]$ value in this processor. Following the circulation the S tokens are routed back to their originating processors.

The algorithm for an SIMD hypercube is given in Program 2.10. f is the SIMD circulation function of Section 2.6. From Theorem 2.1 it follows that $in(p)$ gives the index (within the window) of the PE from which the current S originated. The complexity of this algorithm is O(M). The algorithm for MIMD hypercubes is similar. It uses repeated shifts of -1 to perform the circulation of the S's. Its complexity is also O(M).

```
procedure SIMDConsecutiveSum (X, S, M);
{Consecutive sum of X in windows of size M}
{M is a power of 2}
begin
  in (p) := p mod M; {relative index in size M window}
  S (p) := X [in (p)](p);

  {Circulate S computing the desired sum}
  for i := 1 to M-1 do
  begin
    l := f (log2M, i);
    S (p) <- S (p^(l));
    in (p) := in (p) ⊕ 2^l;
    S (p) := S (p) + X [in (p)];
  end;

  {move S's back to originating PEs}
  j := log2M - 1;
  S (p^(j)) <- S (p);
end; {of SIMDConsecutiveSum}
```

Program 2.10 SIMD consecutive sum

2.9 Adjacent Sum

This operation is defined in Kumar and Krishnan (1987). For each PE, p, $0 \le p < P$, the sum

$$T(p) = \sum_{i=0}^{M-1} A[i]((p+i) \bmod P)$$

is to be computed.

As mentioned earlier, every hypercube of size P can be viewed as consisting of P/M subhypercubes (blocks) each of size M. For every PE p, some (or all) of the A's needed to compute $T(p)$ are in the block containing PE p. The remainder are in the next block of PEs. The strategy to compute T is as follows:

(1) Each PE, p, begins with two variables S and T (initially 0). These values circulate through the M PEs in the block. T accumulates the A values in the block needed in the sum for $T(p)$. S accumulates the A values needed for $T((p-M) \textbf{ mod } P)$.

(2) The S values are shifted clockwise by M positions and added to the T values.

The resulting algorithm for MIMD hypercubes is straightforward. The algorithm for SIMD hypercubes is given in Program 2.11. Recall that at the end of Section 2.6 we extended the definition of $f(p, i)$ to include $f(p, 2^p) = p - 1$. The complexity of the algorithm is $O(M + \log(P/M))$.

2.10 Data Accumulation

For this operation, PE j has an array $A[0..M-1]$ of size M where M is a power of 2. In addition, each PE has a value in its I register. After the data accumulation, the M elements of A in each PE j are such that:

$$A[i](j) = I((j+i) \text{ mod } P),\ 0 \le i < M,\ 0 \le j < P$$

On an MIMD hypercube, data accumulation is done efficiently by performing $M-1$ unit length shifts. On an SIMD hypercube, data accumulation may be done efficiently by adapting the data circulation algorithm of Program 2.9 (Kumar and Krishnan 1987, Ranka and Sahni 1988b). This adaptation takes the form of procedure *SIMDAccum* (Program 2.12).

Following line 3, each block (subhypercube) of size M has all the I values needed to do the data accumulation for that block. These values are in the I_{new} and I_{old} registers of the M PEs in the block. Figure 2.9 (lines

```
procedure SIMDAdjacentSum(A, T, M);
{Adjacent sum in an SIMD hypercube}
begin
  S := 0; T := 0;
  in(p) := p mod M; {in(p) = index of PE p within block of size M}
  for i := 1 to M do {circulate S and T}
  begin
    l := f(log2 M, i);
    S(p) ← S(p^(l));
    T(p) ← T(p^(l));
    in(p) := in(p) ⊕ 2^l;
    T := T + A[p mod M−in(p)], (in(p) ≤ p mod M);
    S := S + A[M+p mod M−in(p)], (in(p) > p mod M);
  end;
  SIMDShift(S,−M,P);
  T := T + S;
end; {of SIMDAdjacentSum}
```

Program 2.11 Adjacent sum

1 and 2) shows the situation for the case $P = 16$ and $M = 8$. In this figure, I_j denotes the value initially in the I register of PE j. The remaining rows give the values of I and l for each of the seven iterations of the **for** loop of lines 6-17.

The variable *HighBit* keeps track of the highest bit encountered in the sequence $f(\log_2 M, i)$ (the function f was defined in Section 2.6). To establish the correctness of procedure *SIMDAccum* we first consider the code fragment of Program 2.13.

From our earlier discussions of f, it follows that $l > HighBit$ exactly when $i = 1, 2, 4, 8, \ldots\ldots$ The operation of the circulation algorithm (Program 2.9), therefore follows the pattern

```
line  procedure SIMDAccum (A, I, M);
      {Each PE accumulates in A, the values of the next M PEs including
      itself}
1       begin
2         I_old := I; I_new := I;
3         SIMDShift (I_new, -M, P);
4         A [0] := I;
5         HighBit := -1;
6         for i := 1 to M-1 do
7         begin
8           l := f (log_2 M, i);
9           I (j) := I_new(j), (l > HighBit) and (j_l = 0);
10          I (j) := I_old(j), (l > HighBit) and (j_l = 1);
11          I (j^(l)) ← I (j);
12          in (j) := j^(l), (l > HighBit) and (j_l = 0);
13          in (j) := j^(l) + M, (l > HighBit) and (j_l = 1);
14          in (j) := in (j) ⊕ 2^l, (l ≤ HighBit);
15          A [(in (j) - j)modP ](j)  := I (j);
16          HighBit := max{l, HighBit}
17        end;
18    end; {of SIMDAccum}
```

Program 2.12 Data accumulation

T ... transfer along $f(\log_2 M, 1)$

T ... transfer along $f(\log_2 M, 2)$

l	PE	0	1	2	3	4	5	6	7	8	9	10	11	12	13	14	15
	I_{old}	I_0	I_1	I_2	I_3	I_4	I_5	I_6	I_7	I_8	I_9	I_{10}	I_{11}	I_{12}	I_{13}	I_{14}	I_{15}
	I_{new}	I_8	I_9	I_{10}	I_{11}	I_{12}	I_{13}	I_{14}	I_{15}	I_0	I_1	I_2	I_3	I_4	I_5	I_6	I_7
0		I_1	I_8	I_3	I_{10}	I_5	I_{12}	I_7	I_{14}	I_9	I_0	I_{11}	I_2	I_{13}	I_4	I_{15}	I_6
1		I_2	I_3	I_8	I_9	I_6	I_7	I_{12}	I_{13}	I_{10}	I_{11}	I_0	I_1	I_{14}	I_{15}	I_4	I_5
0		I_3	I_2	I_9	I_8	I_7	I_6	I_{13}	I_{12}	I_{11}	I_{10}	I_1	I_0	I_{15}	I_{14}	I_5	I_4
2		I_4	I_5	I_6	I_7	I_8	I_9	I_{10}	I_{11}	I_{12}	I_{13}	I_{14}	I_{15}	I_0	I_1	I_2	I_3
0		I_5	I_4	I_7	I_6	I_9	I_8	I_{11}	I_{10}	I_{13}	I_{12}	I_{15}	I_{14}	I_1	I_0	I_3	I_2
1		I_7	I_6	I_5	I_4	I_{11}	I_{10}	I_9	I_8	I_{15}	I_{14}	I_{13}	I_{12}	I_3	I_2	I_1	I_0
0		I_6	I_7	I_4	I_5	I_{10}	I_{11}	I_8	I_9	I_{14}	I_{15}	I_{12}	I_{13}	I_2	I_3	I_0	I_1

Figure 2.9 Data accumulation for $P = 16, M = 8$

```
line
6        for i := 1 to M-1 do
7        begin
8          l := f (logM, i);
9          I(j) := I_new(j), (l > HighBit) and ( j_l = 0);
10         I(j) := I_old(j), (l > HighBit) and ( j_l = 1);
11         I(j^(l)) <- I(j);
16         HighBit := max{l, HighBit}
17       end;
```

Program 2.13 Lines 6-11, 16, and 17 of Program 2.12

C ... circulate independently in hypercubes of size 2
T ... transfer along $f(\log_2 M, 4)$
C ... circulate independently in hypercubes of size 4
T ... transfer along $f(\log_2 M, 8)$
C ... circulate independently in hypercubes of size 8
...
...

This pattern explains the functioning of Program 2.13. When i is a power of 2 ($l > HighBit$), a data transfer along bit $l = f(\log_2 M, i)$ is to be performed. This will be followed by a circulation in subhypercubes of size i (circulation in a subhypercube of size 1 is null). The data required by each subhypercube depends on the l'th bit of the PEs in that subhypercube (all PEs in the subhypercube have the same l'th bit). If this bit is zero, data is coming from a right adjacent subhypercube with l'th bit equal to 1. So, PEs with this bit being 1 should transmit their I_{old} values. On the other hand, subhypercubes with the l'th bit being 1 get data from their left adjacent subhypercube. This has bit $l = 0$ and the required data is in I_{new}. Consequently, the code of Program 2.13 causes the correct values of I to circulate through all the PEs.

Next, we need to establish that lines 12 through 15 of Program 2.12 store the I values in the correct space in A. For this, we first note that when $l > HighBit$, it follows from lines 2, 3, 9, 10, 12, and 13 that $in(j)$ is such that $I(j)$ after the transfer of line 11 originated in processor $in(j)$. The correct place to store this in A is $(in(j) - j)$ **mod** M as is done in line 15. Consequently, line 15 puts the I's in the correct position in A.

It is easily seen that the complexity of procedure *SIMDAccum* is $O(M + \log(P/M))$.

2.11 Rank

Associated with processor, i, in each size 2^k window of a hypercube is a flag $selected(i)$ which is true iff this is a selected processor. The objective of ranking is to assign to each selected processor a *rank* such that $rank(i)$ is the number of selected processors in the window with index less than i. Line 0 of Figure 2.10 shows the selected processors in a window of size eight with an *. An SIMD hypercube is assumed. The ranks to be computed are shown in line 1.

The procedure to compute ranks is very similar to the procedure for prefix sums (Program 2.7). The ranks of the selected processors in a window of size 2^k can be computed easily if we know the following information for the processors in each of the size 2^{k-1} subwindows that comprise the size 2^k window:

(1) Rank of each selected processor in the 2^{k-1} subwindow

(2) Total number of selected processors in each 2^{k-1} subwindow

If a processor is in the left 2^{k-1} subwindow then its rank in the 2^k window is the same as its rank in the subwindow. If it is in the right subwindow, its rank is its rank in the subwindow plus the number of selected processors in the left subwindow. Line 2 of Figure 2.10 shows

					PE				
line	0	1	2	3	4	5	6	7	
0		*	*		*		*	*	
1	∞	0	1	∞	2	∞	3	4	R
2	∞	0	1	∞	0	∞	1	2	R
3	2	2	2	2	3	3	3	3	S
4	0	0	0	0	0	0	0	0	R
5	0	1	1	0	1	0	1	1	S
6	0	0	0	0	0	0	0	1	R
7	1	1	1	1	1	1	2	2	S
8	0	0	1	1	0	0	1	2	R
9	2	2	2	2	3	3	3	3	S
10	0	0	1	1	2	2	3	4	R
11	5	5	5	5	5	5	5	5	S

Figure 2.10 Example to compute ranks in an SIMD hypercube

the rank of each selected processor relative to subwindows of size 4. Line 3 shows the total number of selected processors in each subwindow.

Let $R(i)$ and $S(i)$, respectively, denote the rank of processor i (if it is a selected processor) and the number of selected processors in the current window that contains processor i. Our strategy to compute ranks in windows of size 2^k is to begin with R and S for windows of size one and then repeatedly double the window size until we reach a window size of 2^k. For windows of size one we have:

$$R(i) = 0$$

$$S(i) = \begin{cases} 1 & \text{if } i \text{ is selected} \\ 0 & \text{otherwise} \end{cases}$$

Lines 4 and 5 of Figure 2.10 give the initial R and S values. Lines 6 and 7 give the values for windows of size 2; lines 8 and 9 give these for windows of size 4; and lines 10 and 11 give them for a window size of 8. The ranks for the processors that are not selected may now be set to ∞ to get the configuration of line 1. The procedure to compute ranks is given in Program 2.14. This procedure is due to Nassimi and Sahni (1981) and its complexity is readily seen to be O(k).

```
procedure rank (k);
{Compute the rank of selected processors in windows of size 2^k}
{SIMD hypercube}
begin
  {Initialize for size 1 windows}
  R(i) := 0;
  if selected (i) then S(i) := 1
                  else S(i) := 0;

  {Compute for size 2^(b+1) windows}
  for b := 0 to k-1 do
  begin
    T(i^(b)) <- S(i);
    R(i) := R(i) + T(i), (i_b = 1);
    S(i) := S(i) + T(i);
  end;
  R(i) := ∞, (not selected (i));
end; {of rank}
```

Program 2.14 SIMD ranking procedure

2.12 Concentrate

In a data concentration operation we begin with one record, G, in each of the processors selected for this operation. The selected processors have been ranked and the rank information is in a field R of the record. Assume the window size is 2^k. The objective is to move the ranked records in each window to the processor whose position in the window equals the record rank. Line 0 of Figure 2.11 gives an initial configuration for an SIMD eight processor window. The records are shown as pairs with the second entry in each pair being the rank. We assume that the processors that are not selected for the concentration operation have a rank of ∞. The result of the concentration is shown in line 1.

	PE							
line	0	1	2	3	4	5	6	7
0	(-, ∞)	(B, 0)	(-, ∞)	(D, 1)	(E, 2)	(-, ∞)	(G, 3)	(H, 4)
1	(B, 0)	(D, 1)	(E, 2)	(G, 3)	(H, 4)	(-, ∞)	(-, ∞)	(-, ∞)
2	(B, 0)	(-, ∞)	(-, ∞)	(D, 1)	(E, 2)	(-, ∞)	(H, 4)	(G, 3)
3	(B, 0)	(D, 1)	(-, ∞)	(-, ∞)	(H, 4)	(-, ∞)	(E, 2)	(G, 3)

Figure 2.11 Example to concentrate in an SIMD hypercube

Data concentration can be done in O(k) time by obtaining agreement between the bits of the destination of a record and its present location in the order 0, 1, 2, $\cdots$, $k-1$ (Nassimi and Sahni 1981). For our example, we first seek agreement on bit 0. Examining the initial configuration (line 0) we see that the destination and present location of records B, G, and H disagree on bit 0. To obtain agreement we exchange these records with the records in neighbor processors along bit 0. This gives the configuration of line 2. Examining the bit 1 of destination and present location in line 2 we see that records D, E, and H have a disagreement. Exchanging these records with their neighbors along bit 1 yields line 3. Finally, we examine bit 2 of the destination and present location of records in line 3 and determine that records E and G need to be exchanged with their neighbors along bit 2. This results in the desired final configuration of line 1.

Program 2.15 is the concentration procedure that follows the above strategy. This procedure assumes that the rank information is part of the record and so moves along with the record. Its correctness is not immediate. We need to establish that whenever an exchange is performed, both records involved in the exchange must have disagreement in their destination and present location bits. This is done in Theorem 2.5.

---

procedure *concentrate* (G, k);
{Concentrate records G in selected processors. 2^k is the window size}
{R is the rank field of a record}
begin
  for $b := 0$ **to** $k-1$ **do**
  begin
    $F(i^{(b)}) \leftarrow G(i)$;
    $G(i) \leftarrow F(i)$, $((G(i).R \neq \infty$ **and** $(G(i).R)_b \neq i_b))$
                or $(F(i).R \neq \infty$ **and** $(F(i).R)_b \neq i_b)))$;
  end;
end; {of *concentrate*}

Program 2.15 Procedure to concentrate records

Theorem 2.5 (Nassimi and Sahni 1981) Procedure *concentrate* (Program 2.15) is correct.

Proof: The only condition under which the procedure produces incorrect results is if at the start of some iteration b of the **for** loop we have:

$$G(i).R \neq \infty,\ G(i^{(b)}).R \neq \infty,\ (G(i).R)_b = i_b, \text{ and } ((G(i^{(b)}).R)_b \neq (i^{(b)})_b)$$

for some i. We shall call this the *collision condition*. Suppose that this condition is true. Let j and l, respectively, be the PEs in which the records currently in processors i and $i^{(b)}$ originate. Let $RI(j)$ and $RI(l)$ be the ranks of the records that are initially in these PEs. The collision condition implies that:

$$i = (j_{p-1:b+1} \mid\mid RI(j)_{b:0}) = (l_{p-1:b+1} \mid\mid RI(l)_{b:0})$$

where $\mid\mid$ denotes concatenation, p is the hypercube dimension, and $(u_{w:x} \mid\mid v_{y:z})$ denotes the integer whose binary representation is bits w through x of u concatenated with bits y through z of v.

Since $j \neq l$ and $RI(j) \neq RI(l)$, the equality

$$(j_{p-1:b+1} \mid\mid RI(j)_{b:0}) = (l_{p-1:b+1} \mid\mid RI(l)_{b:0})$$

implies that $\mid j-l \mid < 2^{b+1}$ and $\mid RI(j)-RI(l) \mid \geq 2^{b+1}$. Hence, $\mid j-l \mid < \mid RI(j)-RI(l) \mid$. However, since j and l are in the same 2^k block, $\mid j-l \mid \geq \mid RI(j)-RI(l) \mid$. So, the collision condition cannot hold. □

2.13 Distribute

Data distribution is the inverse of data concentration. We begin with records in processors $0, \cdots, r$ of a hypercube window of size 2^k. Each record has a destination $D(i)$ associated with it. The destinations in each window are such that $D(0) < D(1) < \cdots < D(r)$. The record that is initially in processor i of the window is to be routed to the $D(i)$'th processor of the window. Note that r may vary from window to window. Line 0 of Figure 2.12 gives an initial configuration for data distribution in an eight processor window of an SIMD hypercube. Each record is represented as a tuple with the second entry being the destination. Line 1 gives the result of the distribution.

	PE							
line	0	1	2	3	4	5	6	7
0	(A, 3)	(B, 4)	(C, 7)	(-, ∞)	(-, ∞)	(-, ∞)	(-, ∞)	(-, ∞)
1	(-, ∞)	(-, ∞)	(-, ∞)	(A, 3)	(B, 4)	(-, ∞)	(-, ∞)	(C, 7)
2	(A, 3)	(-, ∞)	(-, ∞)	(-, ∞)	(-, ∞)	(B, 4)	(C, 7)	(-, ∞)
3	(-, ∞)	(-, ∞)	(A, 3)	(-, ∞)	(-, ∞)	(B, 4)	(C, 7)	(-, ∞)

Figure 2.12 Example to distribute in an SIMD hypercube

Since data distribution is the inverse of data concentration it can be carried out by running the concentration procedure in reverse (Nassimi and Sahni 1981). The result is Program 2.16. Lines 2, 3, and 1 of Figure 2.12 give the configurations for our example following the iterations b = 2, 1, and 0, respectively.

```
procedure distribute (G, k);
{Distribute records G. 2^k is the window size}
begin
  for b := k−1 downto 0 do
  begin
    F(i^(b)) ← G(i);
    G(i) ← F(i), ((G(i).D ≠ ∞ and (G(i).D)_b ≠ i_b))
                  or (F(i).D ≠ ∞ and (F(i).D)_b ≠ i_b)));
  end;
end; {of distribute}
```

Program 2.16 Procedure to distribute records

2.14 Generalize

The initial configuration for a generalize is similar to that for a data distribution. We begin with records, G, in processors $0, \cdots, r$ of a hypercube window of size 2^k. Each record, $G(i)$, has a high destination $G(i).H$ associated with it, $0 \le i \le r$. The high destinations in each window are such that $G(0).H < G(1).H < \cdots < G(r).H$. Let $G(-1).H = 0$. The record that is initially in processor i of the window is to be routed to processors $G(i-1).H$, $G(i-1).H + 1$, $\cdots$, $G(i).H$ of the window, $0 \le i \le r$ Note that r may vary from window to window. Line 0 of Figure 2.13 gives an initial configuration for data generalization in an eight processor window of an SIMD hypercube. Each record is represented as a tuple with the second entry being the high destination. Line 1 gives the result of the generalization.

	PE								
line	0	1	2	3	4	5	6	7	
0	(A, 3)	(B, 4)	(C, 7)	(-, ∞)	(-, ∞)	(-, ∞)	(-, ∞)	(-, ∞)	G
1	(A, 3)	(A, 3)	(A, 3)	(A, 3)	(B, 4)	(C, 7)	(C, 7)	(C, 7)	G
2	(-, ∞)	(-, ∞)	(-, ∞)	(-, ∞)	(A, 3)	(B, 4)	(C, 7)	(-, ∞)	F
3	(-, ∞)	(-, ∞)	(-, ∞)	(-, ∞)	(-, ∞)	(B, 4)	(C, 7)	(-, ∞)	F
4	(A, 3)	(B, 4)	(C, 7)	(-, ∞)	(-, ∞)	(B, 4)	(C, 7)	(-, ∞)	G
5	(C, 7)	(-, ∞)	(A, 3)	(B, 4)	(C, 7)	(-, ∞)	(-, ∞)	(B, 4)	F
6	(C, 7)	(-, ∞)	(A, 3)	(B, 4)	(C, 7)	(-, ∞)	(-, ∞)	(-, ∞)	F
7	(A, 3)	(B, 4)	(A, 3)	(B, 4)	(C, 7)	(B, 4)	(C, 7)	(-, ∞)	G
8	(B, 4)	(A, 3)	(B, 4)	(A, 3)	(B, 4)	(C, 7)	(-, ∞)	(C, 7)	F

Figure 2.13 Example to generalize in an SIMD hypercube

Data generalization is done by repeatedly reducing the window size by half (Nassimi and Sahni 1981). Each time the window size is halved we ensure that all records needed in the reduced window are present in that window. Beginning with a window size of eight and line 0 of Figure 2.13, each processor sends its record to its neighbor processor along bit 2. The neighbor processor receives the record in F. Line 2 shows the F values following the transfer. Next, some F's and G's are eliminated. This is done by comparing the high destination of a record with the lowest processor index in the size four window that contains the record. If the compare is true then the record isn't needed in the size four window. Applying this elimination criterion to line 0 results in the elimination of no G. However, when the criterion is applied to the F's of line 2, $F(4)$ is eliminated and we get the configuration of line 3. At this point each window of size four has all the records needed in that window. The records are, however, in both F and G. To consolidate the required records into the G's alone, we use the consolidation criterion:

$$\text{replace } G(i) \text{ by } F(i) \text{ in case } F(i).H < G(i).H$$

I.e., of the two records in a PE, the one with smaller high destination survives. We shall establish the correctness of this consolidation criterion in Theorem 2.6. Applying the consolidation criterion to lines 0 and 3 results in line 4.

Next, records are transferred along bit 1. The F values following this transfer are given in line 5. Following the application of the elimination criterion we get the F values of line 6. The G values are unchanged. When the consolidation criterion is applied the G values are as in line 7. Line 8 shows the F values following a transfer along bit 0. The elimination criterion results in the elimination of no F or G. The consolidation criterion results in line 1. Procedure *generalize* (Program 2.17) implements the generalization strategy just outlined.

Theorem 2.6 (Nassimi and Sahni 1981) Procedure *generalize* is correct.

Proof: For the sake of the proof assume that each record also has a field L

```
procedure generalize (G, k);
{Generalize records G. 2^k is the window size}
begin
  for b := k-1 downto 0 do
  begin
    {Transfer to neighboring window of size 2^b}
    F(i^(b)) <- G(i);

    {Elimination criterion}
    G(i).H := ∞, (G(i).H < i - i_{b-1:0});
    F(i).H := ∞, (F(i).H < i - i_{b-1:0});

    {Consolidation criterion}
    G(i) := F(i), (F(i).H < G(i).H);
  end;
end; {of generalize}
```

Program 2.17 Procedure to generalize records

which gives the index of the lowest PE to which the record is to go. Initially, $L(i) = i$ if i is the lowest index the 2^k block (i.e., $i_{k-1:0} = 0$) and $L(i) = H(i-1) + 1$ otherwise. For line 0 of Figure 2.13 the L values are (0, 4, 5, ∞, ∞, ∞, ∞, ∞). If $H(i) \neq \infty$, then $G(i)$ is to be replicated in PEs $L(i)$ through $H(i)$ of the window.

We define conditions C_1^r and C_2^r which we shall show are true following iteration $b = r, k-1 \leq r \leq 0$, of the **for** loop of procedure *generalize*:

C_1^r Let i and j be any two distinct PE's in the same 2^r block. If $H(i) < L(j)$ then $L(j) - R(i) \geq (j-i) \bmod 2^r)$

C_2^r The G's in each 2^r block contain at least one copy of each record needed in that block

The theorem will then follow from the truth of C_2^0. The proof is by induction of r. C_1^k and C_2^k are easily seen to be true. Assume that C_1^r and C_2^r are true for $r = w+1$. We shall show that C_1^w and C_2^w are also true. First consider C_1^w. If $w = 0$, then C_1^0 is trivially true. So, assume that $w > 0$. Let i and j, $i \neq j$ be the indices of two PE's that are in the same 2^w block. Let $G(i)$ and $G(j)$, respectively, be the records in these two PE's at the end of the iteration $b = w$. Let l and u, respectively, be their locations at the start of this iteration (and hence at the end of the previous iteration (if any)). If $H(j) < L(i)$ at the end of iteration $b = w$, then $H(l) < L(u)$ at the start of the iteration. From the truth of C_1^{w+1}, we get:

$$L(u) - H(l) \geq (u - l) \bmod 2^{w+1}$$

Using $H(l) = H(i)$ and $L(u) = L(j)$ in this inequality, we get:

$$L(j) - H(i) \geq (u - l) \bmod 2^{w+1}$$

Since, $l \in \{i, i+2^w, i-2^w\}$ and $u \in \{j, j+2^w, j-2^w\}$,

$$(u - l) \bmod 2^{w+1} \geq (j - i) \bmod 2^w)$$

Hence C_1^w is true following the iteration $b = w$.

Next consider C_2^w. From C_2^{w+1} and the transfer to neighbor step, it follows that after the transfer to neighbor step is complete the G's and F's in each 2^w window collectively contain all the records needed in the 2^w window. Since the elimination criterion only eliminates records that are not needed in the window, we need be concerned only about the consolidation step. Furthermore, we need only concern ourselves with processors i for which $H.F(i) \neq \infty$ and $H.G(i) \neq \infty$. If $H.G(i) < H.F(i)$, then from C_1^{w+1} and the fact that $G(i)$ and $F(i)$ were 2^w apart before the transfer we get:

$$L.F(i) - H.G(i) \geq 2^w \bmod 2^{w+1} = 2^w$$

This and the observation that $H.G(i) \geq i - i_{w-1:0}$ implies that $L.F(i) \geq$

$i - i_{w-1:0} + 2^w$. Hence, $L.F(i)$ is greater than any PE index in the 2^w window that contains i. So, $F(i)$ is not needed in the 2^w block and may be eliminated. A similar argument shows that when $H.F(i) < H.G(i)$, $G(i)$ is not needed. When $H.F(i) = H.G(i)$, $F(i)$ and $G(i)$ are the same record and it doesn't matter which is eliminated. So, following the consolidation step C_2^w is true. □

2.15 Sorting

A *bitonic sequence* is a nonincreasing sequence of numbers followed by a nondecreasing sequence. Either (or both) of these may be empty. The sequence has the form $x_1 \geq x_2 \geq \cdots \geq x_k \leq x_{k+1} \leq \cdots \leq x_n$, for some k, $1 \leq k \leq n$. The sequences 10, 9, 9, 4, 5, 7, 9; 2, 3, 4, 5, 8; 7, 6, 4, 3, 1; and 11, 2, 5, 6, 8, 9 are example bitonic sequences.

A *bitonic sort* is a process that sorts a bitonic sequence into either nonincreasing or nondecreasing order. A bitonic sort can be used to merge together two sorted sequences $v_1 \leq v_2 \leq \cdots \leq v_l$ and $w_1 \leq w_2 \leq \cdots \leq w_m$ by first concatenating them to obtain the bitonic sequence $v_l \geq v_{l-1} \geq \cdots \geq v_1$? $w_1 \leq w_2 \leq \cdots \leq w_m = x_1 \geq x_2 \geq \cdots \geq x_k \leq x_{k+1} \leq \cdots \leq x_n$ where $n = l + m$. The resulting bitonic sequence is then sorted using a bitonic sort to obtain the desired merged sequence. So, for example, if we wish to merge the sequences (2, 8, 20, 24) and (1, 9, 10, 11, 12, 13, 30) we first create the bitonic sequence (24, 20, 8, 2, 1, 9, 10, 11, 12, 13, 30).

Batcher's bitonic sort (Knuth 1973) is ideally suited for implementation on a hypercube computer. Batcher's algorithm to sort the bitonic sequence $x_1, \cdots, x_n$ into nondecreasing order is given in Program 2.18.

Example 2.1 Consider the bitonic sequence (24, 20, 8, 2, 1, 9, 10, 11, 12, 13, 30). Suppose we wish to sort this into nondecreasing order. The odd sequence is (24, 8, 1, 10, 12, 30) and the even sequence is (20, 2, 9, 11, 13). Sorting these, we obtain the sequences (1, 8, 10, 12, 24, 30) and (2, 9, 11, 13, 20). Putting the sorted odd and even parts together, we obtain the sequence (1, 2, 8, 9, 10, 11, 12, 13, 24, 20, 30). After performing the

---

Step 1: [Sort odd subsequence] If $n > 2$ then recursively sort the odd bitonic subsequence $x_1, x_3, x_5, \cdots$ into nondecreasing order

Step 2: [Sort even subsequence] If $n > 3$ then recursively sort the even bitonic subsequence $x_2, x_4, x_6, \cdots$ into nondecreasing order

Step 3: [Compare/exchange] Compare the pairs of elements x_i and x_{i+1} for i odd and exchange them in case $x_i > x_{i+1}$

Program 2.18 Bitonic sort into nondecreasing order

$\lfloor n/2 \rfloor$ compare/exchanges of step 3, we obtain the sorted sequence (1, 2, 8, 9, 10, 11, 12, 13, 20, 24, 30). □

The 0/1 principle may be used to establish the correctness of Batcher's method.

Theorem 2.7 [0/1 Principle] (Knuth 1973) If a sorting algorithm that performs only element comparisons and exchanges sorts all sequences of zeroes and ones then it sorts all sequences of arbitrary numbers.

Proof: We shall show that if a compare exchange algorithm fails to sort a single sequence of arbitrary numbers then it must fail to sort at least one sequence of zeroes and ones. Hence if all 0/1 sequences are sorted then all arbitrary sequences are also sorted.

Let f be any monotonic function such that $f(x) \leq f(y)$ whenever $x \leq y$. It is easy to see that if a compare/exchange algorithm transforms $(x_1, \cdots, x_n)$ into $(y_1, \cdots, y_n)$, then it transforms $(f(x_1), \cdots, f(x_n))$ into $(f(y_1), \cdots, f(y_n))$.

Suppose that the algorithm transforms $(x_1, \cdots, x_n)$ into $(y_1, \cdots, y_n)$ and $y_i > y_{i+1}$ for some i (i.e., the input sequence isn't sorted). Define the monotonic function f such that $f(x_j) = 0$ for $x_j < y_i$ and $f(x_j) = 1$ for $x_j \geq y_i$. The algorithm transforms the 0/1 sequence $(f(x_1), \cdots, f(x_n))$ into the sequence $(f(y_1), \cdots, f(y_n))$ which is not sorted. □

As a result of Theorem 2.7 the correctness of Program 2.18 can be established by showing that this algorithm sorts all 0/1 bitonic sequences.

Theorem 2.8 (Knuth 1973) Program 2.18 sorts all 0/1 bitonic sequences.

Proof: We shall use induction on the length n of the 0/1 bitonic sequence. The correctness of Program 2.18 is easily verified for $n \leq 2$. Assume its correctness for $n \leq m$ where m is an arbitrary natural number greater than 1. Consider any 0/1 bitonic sequence $x_1, \cdots, x_n$ with $n = m + 1$. Its odd and even subsequences are bitonic sequences of length less than $m+1$. From the induction hypothesis Program 2.18 correctly sorts these. The initial 0/1 bitonic sequence is of the form $1^a 0^b 1^c$ where q^r denotes a sequence of r q's and $a + b + c = n = m + 1$. Hence if the odd subsequence contains d zeroes and if the even subsequence contains e zeroes, then $| d - e | \leq 1$. The sorted odd subsequence has the form $0^d 1^{\lceil n/2 \rceil - d}$ and the sorted even subsequence has the form $0^e 1^{\lfloor n/2 \rfloor - e}$. If $d = e$ or $d = e + 1$ then the combination of the sorted odd and even subsequences is also sorted and no exchanges take place in step 3. If $d = e - 1$, then the combination of the sorted odd and even subsequences has a 0 in position $2e$ and a 1 in position $2e-1$. The elements in positions 1 through $2e - 2$ are all zeroes and those in positions $e + 1$ through n are all ones. The elements in positions $2e - 1$ and $2e$ are compared in step 3 and exchanged. Hence following step 3 we have a sorted sequence. □

When n is a power of 2 the recursion of Program 2.18 can be unfolded to obtain the compare/exchange algorithm of Program 2.19. In each iteration of the **while** loop each sequence element is paired with exactly one other sequence element that is a distance d from it. The pairs are formed left to right. To obtain a nondecreasing sequence each compare exchange causes the smaller element of the pair to move to the left position. If a nonincreasing sequence is desired the smaller element is moved to the right. Figure 2.14 shows an eight element bitonic merge that results in a nondecreasing sequence and Figure 2.15 gives an example that results in a nonincreasing sequence. The examples assume the elements to be sorted are stored in processors of a hypercube with one element per processor. As can be seen the elements that form each of the pairs for the compare/exchange operation are in processors that are hypercube neighbors. Hence each iteration of the **while** loop of Program 2.19 takes O(1) time on a hypercube. The total time to sort an n element bitonic sequence is therefore O(log$n$).

```
procedure BitonicSort(n);
{Sort the bitonic sequence x_1, ..., x_n}
{n is a power of 2}
begin
  d = n/2;
  while d > 0 do
  begin
    compare/exchange elements d apart
    d = d/2;
  end;
end; {of BitonicSort}
```

Program 2.19 Iterative bitonic sort for n a power of 2

To sort n elements using bitonic sort, we begin with sorted sequences of size one. Adjacent pairs of these form bitonic sequences that are sorted (in parallel) to obtain sorted sequences of size two. The sort is done such that the size two sequences are alternately nonincreasing and nondecreasing sequences (i.e, the first, third, fifth, ..., sequences are nonincreasing and the remainder are nondecreasing). Consequently every pair of adjacent size two sequences forms a bitonic sequence of size four which can be sorted using bitonic sort. The size four sequences are also sorted alternately into nonincreasing and nondecreasing order. Continuing in this way we can obtain a sorted sequence of size n after $\log_2 n$ bitonic sort steps. Note that if the sorted sequence is to be in nondecreasing order, then the last bitonic sort step should sort the first and only resulting sequence into this order. The total time for the sort is

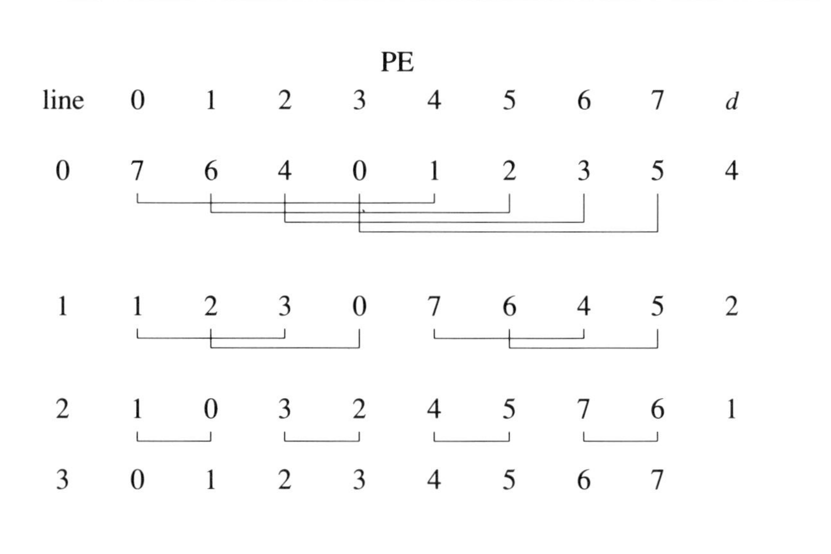

Figure 2.14 Power of 2 bitonic sort (nondecreasing order)

$O(\log^2 n)$.

Example 2.2 Suppose we wish to sort the sequence

c n m f h a p d g j l k b e i o

into nondecreasing order and that a < b < $\cdots$ < o < p. The pairs (c n), (m f), (h a), (p d), (g j), (l k), (b e), and (i o) are bitonic sequences that are sorted using bitonic sort to obtain the sequence:

n c f m h a d p j g k l e b i o

Note that the odd pairs were sorted into nonincreasing order while the even ones were sorted into nondecreasing order. Next we consider adjacent sequences of length four. These are (n c f m), (h a d p), (j g k l), and (e b i o). Since each is a bitonic sequence it may be sorted using bitonic sort. The result is:

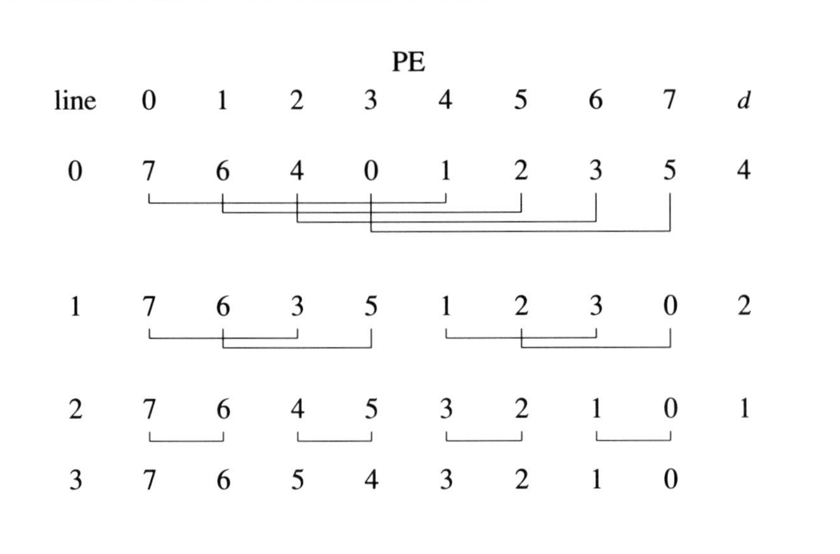

Figure 2.15 Power of 2 bitonic sort (nonincreasing order)

n m f c a d h p l k j g b e i o

Once again the odd sequences are sorted into nonincreasing order while the even ones are sorted into nondecreasing order. We now have two bitonic sequences of length eight: (n m f c a d h p) and (l k j g b e i o). Sorting these gives us the sequence:

p n m h f d c a b e g i j k l o

Sorting this into nondecreasing order results in the sequence:

a b c d e f g h i j k l m n o p

□

2.16 Random Access Read

In a random access read (RAR) some of the processors of the hypercube wish to read data from other processors of the hypercube. Let $A(i)$ be the PE from which processor i wishes to get data. The data to be obtained is $D(A(i))$. In case PE i does not wish to read data from any other PE, then $A(i) = \infty$. Line 0 of Figure 2.16 gives the A values for an example RAR in an eight processor hypercube. Note that in an RAR several processors may read from the same PE. An RAR can be done in $O(\log^2 n)$ time in an n processor hypercube using the algorithm of Program 2.20 (Nassimi and Sahni 1981).

Step 1: Each processor a creates a triple $(A(a), a, flag)$ where $flag$ is a Boolean entity that is initially true.

Step 2: [Sort] Sort the triples into nondecreasing order of the read address $A(a)$. Triples with the same read address are in nondecreasing order of the PE index a. Furthermore, during the sort the $flag$ entry of a triple is set to false in case there is a triple to its right with the same read address.

Step 3: [Rank] Processors with triples whose first component $\neq \infty$ and whose third component (i.e., $flag$) is true are ranked.

Step 4: Each processor b that has a triple $(A(a), a, \text{true})$ with $A(a) \neq \infty$ creates a triple of the form $(R(b), A(a), b)$ where $R(b)$ is the rank computed in the preceding step.

Step 5: [Concentrate] The triples just created are concentrated.

Step 6: Each processor c that has a concentrated triple $(R(b), A(a), b)$ creates a tuple of the form $(c, A(a))$. Note that since $c = R(b)$ this tuple is just the first two components of the triple.

Step 7: [Distribute] The tuples are distributed using the second component as the destination address.

Step 8: Each processor $A(a)$ that receives a tuple $(c, A(a))$ creates the tuple $(c, D(A(a)))$.

	PE								
step	0	1	2	3	4	5	6	7	
0	4	3	∞	1	7	7	∞	3	A
1	(4, 0, t)	(3, 1, t)	(∞, 2, t)	(1, 3, t)	(7, 4, t)	(7, 5, t)	(∞, 6, t)	(3, 7, t)	
2	(1, 3, t)	(3, 1, f)	(3, 7, t)	(4, 0, t)	(7, 4, f)	(7, 5, t)	(∞, 2, f)	(∞, 6, t)	sort
3	0		1	2		3			rank
4	(0, 1, 0)		(1, 3, 2)	(2, 4, 3)		(3, 7, 5)			
5	(0, 1, 0)	(1, 3, 2)	(2, 4, 3)	(3, 7, 5)					concentrate
6	(0, 1)	(1, 3)	(2, 4)	(3, 7)					
7		(0, 1)		(1, 3)	(2, 4)			(3, 7)	distribute
8		(0, $D(1)$)		(1, $D(3)$)	(2, $D(4)$)			(3, $D(7)$)	
9	(0, $D(1)$)	(1, $D(3)$)	(2, $D(4)$)	(3, $D(7)$)					concentrate
10	(0, $D(1)$)	(2, $D(3)$)	(3, $D(4)$)	(5, $D(7)$)					
11	(0, $D(1)$)	(2, $D(3)$)	(2, $D(3)$)	(3, $D(4)$)	(5, $D(7)$)	(5, $D(7)$)			generalize
12	(3, $D(1)$)	(1, $D(3)$)	(7, $D(3)$)	(0, $D(4)$)	(4, $D(7)$)	(5, $D(7)$)	(2, -)	(6, -)	
13	(0, $D(4)$)	(1, $D(3)$)	(2, -)	(3, $D(1)$)	(4, $D(7)$)	(5, $D(7)$)	(6, -)	(7, $D(3)$)	sort

Figure 2.16 Example of a random access read

Step 9: [Concentrate] The tuples created in the preceding step are concentrated using the first component as the rank.

Step 10: Each processor c that received a tuple $(c, D(A(a)))$ in the last step also has a triple of the form $(R(b), A(a), b)$ that it received in Step 5 (notice that $c = R(b)$). Using this triple and the tuple received in Step 9 it creates the tuple $(b, D(A(a)))$.

Step 11: [Generalize] The tuples $(b, D(A(a)))$ are generalized using the first component as the high destination.

Step 12: Each processor that received a tuple $(b, D(A(a)))$ in Step 11 also has a triple $(A(a), a, \mathit{flag})$ that it obtained as a result of the sort of Step 2. Using information from the tuple and the triple it creates a new tuple $(a, D(A(a)))$. Processors that did not receive a tuple use the triple they received in Step 2 and form the tuple $(a, \text{-})$.

Step 13: [Sort] The newly created tuples of Step 12 are sorted by their first component.

Program 2.20 Algorithm for a random access read

Consider the example of Figure 2.16. In Step 1 each processor creates a triple with the first component being the index of the processor from which it wants to read data; the second component is its own index; and the third component is a flag that is initially true (t) for all triples. Then, in Step 2 the triples are sorted on the first component. Triples that have the same first component are in increasing order of their second component. Within each sequence of triples that have the same first component only the last one has a true flag. The flag for the remaining triples is false. The first components of the triples with a true flag give all the distinct processors from which data is to be read.

Processors 0, 2, 3, and 5 are ranked in Step 3. Since the highest rank is three, data is to read from only four distinct processors. In Step 4 the ranked processors create triples of the form $(R(b), A(a), b)$. The triples are then concentrated. Processors 0 through 3 receive the concentrated triples and form tuples of the form $(c, A(a))$. Because of the sort of

Step 2 the second components of these tuples are in ascending order. Hence, they can be routed to the processors given by the second component using a data distribution as in Step 7. The destination processors of these tuples are the distinct processors whose data is to be read. These destination processors create, in Step 8, tuples of the form $(c, D(A(a)))$ where c is the index of the processor that originated the tuple it received. These tuples are concentrated in Step 9 using the first component as the rank.

In Step 10 the receiving processors (i.e., 0 through 3) use the triples received in Step 5 and the tuples received in Step 9 to create tuples of the form $(b, D(A(a)))$. The first component is the index of the processor that originated the triple received in Step 5. Since the triples received in Step 5 are the result of a concentration, the first component of the newly formed tuples are in ascending order. The tuples are therefore ready for generalization using the first component as the high index. This is done in Step 11. After this generalization we have the right number of copies of each data. For example, two processors (4 and 5) wanted to read from processor 7 and we now have two copies of $D(7)$. Comparing the triples of Step 2 and the tuples of Step 11, we see that the second component of the triples tells us where the data in the tuples is to be routed to. In Step 12 we create tuples that contain the destination processor and the data. Since the destination addresses are not in ascending order the tuples cannot be routed to their destination processors using a distribute. Rather, they must be sorted by destination.

2.17 Random Access Write

A random access write (RAW) is like a random access read except that processors wish to write to other processors rather than to read from them. A random access write uses many of the basic steps used by a random access read. It is, however, quite a bit simpler. Line 0 of Figure 2.17 gives the index $A(i)$ of the processor to which processor i wants to write its data $D(i)$. $A(i) = \infty$ in case processor i is not to write to another processor. Observe that it is possible for several processors to have the same write address A. When this happens, we say that the RAW has collisions. It is possible to formulate several strategies to handle collisions. Three of these are:

(1) [Arbitrary RAW] Of all the processors that attempt to write to the same processor exactly one succeeds. Any of these writing processors may succeed.

(2) [Highest/lowest RAW] Of all the processors that attempt to write to the same processor the one with the highest (lowest) index succeeds.

(3) [Combining RAW] All the porcessors succeed in getting their data to the target processors.

Consider the example of line 0 of Figure 2.17. In an arbitrary RAW any one of $D(0)$, $D(2)$, and $D(7)$ will get to processor 3; one of $D(1)$ and $D(5)$ will get to processor 0; and $D(3)$ and $D(4)$ will get to processors 4 and 6, respectively. In a highest RAW $D(7)$, $D(5)$, $D(3)$, and $D(4)$, respectively, get to processors 3, 0, 4, and 6. In a lowest RAW $D(0)$, $D(1)$, $D(3)$ and $D(4)$ get to processors 3, 0, 4, and 6, respectively. In a combining RAW $D(0)$, $D(2)$, and $D(7)$ all get to processor 3; both $D(1)$ and $D(5)$ get to processor 0; and $D(3)$ and $D(4)$ get to processors 4 and 6, respectively.

The steps involved in an arbitrary RAW are given in Program 2.21 (Nassimi and Sahni 1981). Let us go through these steps on the example of Figure 2.17. Each processor first creates triples whose first component is the index, $A(a)$, of the processor to which it is to write. Its second component is the data, $D(a)$, to be written and the third component is true. The triples are then sorted on the first component. During this sort the flag entry of a triple is changed to false in case there is a triple with the same write address to its right. Only the triples with a true flag are invloved in the remainder of the algorithm. Notice that for each distinct write address there will be exactly one triple with a true flag. The processors that have a triple with a true flag are ranked (Step 3) and these processors create new triples whose first and second components are the same as in the old triples but whose third component is the rank. The triples are then concentrated using this rank information. Since the triples

	PE								
step	0	1	2	3	4	5	6	7	
0	3	0	3	4	6	0	∞	3	A
1	(3,D(0),t)	(0,D(1),t)	(3,D(2),t)	(4,D(3),t)	(6,D(4),t)	(0,D(5),t)	(∞,D(6),t)	(3,D(7),t)	
2	(0,D(1),f)	(0,D(5),t)	(3,D(0),f)	(3,D(2),f)	(3,D(7),t)	(4,D(3),t)	(6,D(4),t)	(∞,D(6),t)	sort
3		0			1	2	3		rank
4		(0,D(5),0)			(3,D(7),1)	(4,D(3),2)	(6,D(4),3)		
5	(0,D(5),0)	(3,D(7),1)	(4,D(3),2)	(6,D(4),3)					concentrate
6	(0,D(5),0)			(3,D(7),1)	(4,D(3),2)		(6,D(4),3)		distribute

Figure 2.17 Example of an arbitrary random access write

are in ascending order of the write addresses (first component) they may be routed to these processors using a data distribute operation. Note that for Step 6 the third component (i.e., rank) of each triple may be dropped before the distribute begins.

The complexity of a random access write is determined by the sort step which takes $O(\log^2 n)$ time where n is the number of processors.

Step 1: Each processor a creates a triple $(A(a), D(a), flag)$ where *flag* is a Boolean entity that is initially true.

Step 2: [Sort] Sort the triples into nondecreasing order of the write address $A(a)$. Ties are broken arbitrarily and during the sort the *flag* entry of a triple is set to false in case there is a triple to its right with the same write address.

Step 3: [Rank] Processors with triples whose first component is not ∞ and whose third component (i.e., *flag*) is true are ranked.

Step 4: Each processor b that has a triple $(A(a), D(a)$, true) with $A(a) \neq \infty$ creates a triple of the form $(A(a), D(a), R(b))$ where $R(b)$ is the rank computed in the preceding step.

Step 5: [Concentrate] The triples just created are concentrated.

Step 6: [Distribute] The concentrated triples are distributed using the first component as the destination address.

Program 2.21 Algorithm for an arbitrary RAW

A highest (lowest) RAW can be done by modifying Program 2.21 slightly. Step 1 creates 4-tuples instead of triples. The fourth component is the index of the originating processor. In the sort step (Step 2) ties are broken by the fourth component in such a way that the right most 4-tuple in any sequence with the same write address is the 4-tuple we want to succeed (i.e., highest or lowest fourth component in the sequence). Following this the fourth component may be dropped from each 4-tuple.

The remaining steps are unchanged.

The steps for a combining RAW are also similar to those in Program 2.21. When the ranking of Step 3 is done we use a version of procedure *rank* (Program 2.14) that does not contain the last line ($R(i) := \infty$, (**not** *selected* (i))). As a result processor 0 (Figure 2.17) has a rank of 0 and processors 2 and 3 have a rank of 1. During the concentration step (Step 5) more than one triple will try to get to the same processor. Procedure *concentrate* (Program 2.15) is modified to combine together triples that have the same rank. These modifications do not change the asymptotic complexity of the RAW unless the combining operation increases the triple size (as in a concatenate). In case d data values are to reach the same destination, the complexity is $O(\log^2 n + d\log n)$.

2.18 BPC Permutations

The sorting problem of Section 2.15 requires us to rearrange the records in nondescending order of key value. The desired rearrangement is simply a permutation of the initial order and as we saw in Section 2.15, this permutation can be performed on a hypercube in $O(\log^2 n)$ time where n is both the number of records and the number of processors in the hypercube. In several situations where the records are to be permuted, the desired permutation can be specified by providing an explicit relation between a records initial and final positions. One class of permutations obtained by such an explicit specification is known as the bit-permute-complement (BPC) class (Nassimi and Sahni 1982). In this class, the final or destination PE of the record initially in PE i is obtained by permuting and possibly complementing some of the bits in the binary representation of i.

Every BPC permutation on a hypercube of dimension k (i.e., $n = 2^k$) can be specified by a permutation vector $B = [B_{k-1}, B_{k-2}, \ldots, B_0]$ such that $[\mid B_{k-1} \mid, \mid B_{k-2} \mid, \ldots, \mid B_0 \mid]$ is a permutation of $[k-1, k-2, \ldots, 0]$. In the specification of B, we distinguish between -0 and 0. Thus, -0 is considered to be less than 0. Let $d = d_{k-1}d_{k-2}\ldots d_0$ be the destination PE of the data initially in PE $i = i_{k-1}i_{k-2}\ldots i_0$. i_j determines bit $|B_j|$ of d as below:

$$d_{|B_j|} = \begin{cases} i_j & \text{if } B_j \geq 0 \\ \overline{i_j} & \text{if } B_j < 0 \end{cases}$$

where $\overline{i_j}$ is the complement of i_j.

As an example, consider the case $k = 3$ and $B = [0, 2, 1]$. The binary representation of the destination d of the data in PE $i = i_2 i_1 i_0$ is $i_1 i_0 i_2$. The mapping from initial to destination PEs is as given in Figure 2.18. Figure 2.19 gives the mapping for the case $B = [1, \overline{0}, 2]$. In this case $d = i_0 \overline{i_2} i_1$.

i	$i_2 i_1 i_0$	$d_2 d_1 d_0$	d
0	000	000	0
1	001	010	2
2	010	100	4
3	011	110	6
4	100	001	1
5	101	011	3
6	110	101	5
7	111	111	7

Figure 2.18 $B = [0, 2, 1]$

Some of the commonly performed BPC permutations and their corresponding B vectors are given in Figure 2.20. Notice that the example $B = [0, 2, 1]$ considered above is an example of a perfect shuffle permutation.

Every BPC permutation can be performed $O(\log^2 n)$ time by having each PE compute the destination PE for its data and then sorting the data using the destination PE as key. It is possible to perform BPC

i	$i_2 i_1 i_0$	$d_2 d_1 d_0$	d
0	000	001	1
1	001	101	5
2	010	000	0
3	011	100	4
4	100	011	3
5	101	111	7
6	110	010	2
7	111	110	6

Figure 2.19 $B = [1, \bar{0}, 2]$

Permutation	B
Matrix Transpose	$[k/2-1, \ldots, 0, k-1, \ldots, k/2]$
Bit Reversal	$[0, 1, 2, \ldots, k-1]$
Vector Reversal	$[-(k-1), -(k-2), \ldots, -0]$
Perfect Shuffle	$[0, k-1, k-2, \ldots, 1]$
Unshuffle	$[k-2, k-3, \ldots, 0, k-1]$
Shuffled Row Major	$[k-1, k/2-1, k-2, k/2-2, \ldots, k/2, 0]$
Bit Shuffle	$[k-1, k-3, \ldots, 1, k-2, k-4, \ldots, 0]$

Figure 2.20 Common BPC permutations (Nassimi and Sahni 1982)

permutations in O(log$n$) time by an algorithm tailored to these permutations.

The strategy of the algorithm of Nassimi and Sahni (1982) is to route data along the hypercube dimensions such that following the route along dimension b, the present location of data and its final destination agree on bit b, $0 \le b < k$. The order in which the bits/dimensions for routing are selected is based on the cycle structure of the permutation B. The permutation B is examined right to left (i.e., in the order $B_0, B_1, \ldots, B_{k-1}$). If $B_b = b$, then the present and final location of the data agree on bit b and no routing on this bit is to be performed. If $B_b = \bar{b}$, then the final and present location of all data differ on bit b. To get agreement, it is necessary to route all data along bit b. This results in data in every PE i with $i_b = 0$ being moved to the corresponding PE with bit $b = 1$ and vice versa. Agreement in bit b of the present and final locations is obtained and at the same time agreement in the other bits is unaffected. In case $|B_b| \ne b$, a nontrivial cycle (i.e., one of length greater than 1) begins at bit b. The next position on this cycle is $c = |B_b|$. Let $d = |B_c|$. Note that $d \ne c$. If $d = b$, then c is the end of the cycle. If $d \ne b$, then d is the next position in the cycle. Following in this way, a complete cycle of B can be identified.

Consider the perfect shuffle permutation given by $B = [B_2, B_1, B_0] = [0, 2, 1]$ (Figure 2.18). This consists of the single cycle $b = 0$, $c = 1$, and $d = 2$. We shall refer to this as the cycle 0, 1, 2. The permutation $B = [4, -6, 0, -7, 2, -5, 1, 3]$ consists of the cycles 0, 3, 2, 5; 1; 4, 7; and 6. The cycles 1 and 6 are of length 1 and are called trivial cycles. The remaining cycles are nontrivial cycles. The BPC permutation algorithm handles these cycles in the order listed. When the cycle 0, 3, 2, 5 is handled, we obtain agreement between the present and final locations of data on their 0, 3, 2, and 5 bits. Next, the trivial cycle 1 is handled. This requires no routing as the present and final locations of data already agree on bit 1. Handling of the cycle 4, 7 requires us to obtain agreement on bits 4 and 7 of the present and final locations of all data. Finally, the cycle 6 is handled. Since $B_6 = -6$, the present and final locations of all data disagree on this bit. Agreement is obtained by routing all data along dimension 6 of the hypercube.

To handle a nontrivial cycle, assume that the data to be rearranged is in $R(0:n-1)$. We shall make use of the register $S(i)$ in PE i, $0 \le i < n$ to hold data that is to be routed along the next routing dimension. When considering the cycle $c_0, c_1, c_2, \ldots$, we shall first route along dimension c_1 to obtain agreement on bit c_1; the next route is along dimension c_2 and results in agreement on bit c_2; ...; the final route is along dimension c_0 and obtains agreement on bit c_0. For the cycle 0, 1, 2, the routing sequence is dimension 1 first, then dimension 2, and finally dimension 0. The routing sequence for the cycle 0, 3, 2, 5 is 3, 2, 5, 0. The case of the cycle 0, 1, 2 is worked out in Figure 2.21.

In Figure 2.21, the column labeled PE gives the binary representation of the eight PE indices. $OR(i)$ and $OS(i)$, respectively give the index of the originating PE of the data currently in $R(i)$ and $S(i)$. The symbol $*$ is used when $R(i)$ or $S(i)$ contains no data. Column 1 gives the initial data configuration. The first route will be along dimension one of the hypercube and is to result in agreement on bit one. From Figure 2.18, we see that the present and final locations of $R(0)$, $R(3)$, $R(4)$, and $R(7)$ already agree on bit one. So, these are not to be routed along dimension one. Bit one of the present location of the remaining records does not agree with bit one of their destinations. These records are to be routed along dimension one. For this, they are first moved to the corresponding S registers. This movement is actually done by exchanging the R and S values in the corresponding PEs. The exchange pairs are identified by double headed arrows in Figure 2.21. The result is column 2 of Figure 2.18. Routing S along dimension one results in exchanging the S pairs identified by the double headed arrows of column 2. Column 3 gives the configuration following this route. Now the current and final locations of all records agree on bit one. Observe that half the records are in R and the remainder are in S.

The next routing dimension is dimension two. Since bit one of an originating index is to become bit two of the destination index, we need to route $R(i)$ if $(OR(i))_1 \neq (OR(i))_2$ and $S(i)$ is to be routed if $(OS(i))_1 \neq (OS(i))_2$. Since there has been no routing on bit two thus far, $(OR(i))_2$ and $(OS(i))_2$ both equal i_2. Further, since $S(i)$ was just routed on bit one and $R(i)$ has not been routed on bit one, $(OS(i))_1 = \bar{i}_1$ and $(OR(i))_1 = i_1$. Consequently, $S(i)$ is to be routed if $i_1 = i_2$ and $R(i)$ is to be routed otherwise. To prepare for this routing, the data to be routed is moved to S in case it isn't already there. This requires us to exchange $R(i)$ and $S(i)$ in all PEs with $i_1 \neq i_2$. That is, PEs 2, 3, 4, and 5 exchange their R and S values.

PE	Column 1	2	3	4	5	6	7	8	
000	0	0	0	0	0	0	0	0	*OR*
	*	*	2	2	4	4	*	–	*OS*
001	1	*	*	*	*	*	*	4	*OR*
	*	1	*	*	*	*	4	–	*OS*
010	2	*	*	*	*	*	*	1	*OR*
	*	2	*	*	*	*	1	–	*OS*
011	3	3	3	1	1	5	5	5	*OR*
	*	*	1	3	5	1	*	–	*OS*
100	4	4	4	6	6	2	2	2	*OR*
	*	*	6	4	2	6	*	–	*OS*
101	5	*	*	*	*	*	*	6	*OR*
	*	5	*	*	*	*	6	–	*OS*
110	6	*	*	*	*	*	*	3	*OR*
	*	6	*	*	*	*	3	–	*OS*
111	7	7	7	7	7	7	7	7	*OR*
	*	*	5	5	3	3	*	–	*OS*
Route dimension		1		2		0			

Figure 2.21 Perfect shuffle

These exchanges are identified by the double headed arrows of column 3. The result is shown in column four. Next, all PEs route their S values along dimension two. Column four identifies the processor pairs involved in the exchange. Only pairs with useful S data are marked. The configuration following the dimension two routing is shown in column 5.

The last route is along dimension zero. Column 5 shows the processors that need to exchange their R and S values so that data that is to be routed along dimension zero is in the S registers of all PEs. Column 6 shows the result of the exchange and column seven shows the configuration following the routing along dimension zero. Comparing with Figure 2.18, we see that all records have been successfully routed to their correct destination PEs. All that remains is for the four PEs that have the records in their S registers to move them into their R registers. The final configuration is shown in column eight.

The preceding discussion may be generalized to arrive at the algorithm of Program 2.22. The symbol :=: denotes an exchange and $a :=: b$ is equivalent to the statements $t := a$; $a := b$; $b := a$. To establish the correctness of this procedure, we need to show the following:

(1) Lines 10 and 11 correctly set the S registers to contain the data that is to be routed along dimension q.

(2) Lines 18 and 19 correctly move records back to the R registers.

This is done in the next two theorems.

Theorem 2.9 Let $DR(i)$ and $DS(i)$, respectively, denote the destination PE of the data currently in $R(i)$ and $S(i)$. In case $OR(i)$ ($OS(i)$) equals $*$, let $DR(i)$ ($DS(i)$) equal $*$. Assume that all Boolean expressions involving $*$ are true. So, $*_j = i_j$ and $*_j = \overline{i_j}$ are both true. Following line 11 of Program 2.22, we have $(DR(i))_q = i_q$ and $(DS(i))_q \neq i_q$, $0 \leq i < n$. So, exactly the S register data is to routed along dimension q in order to get bit q of the current and destination locations to be the same.

Proof: At the start of each nontrivial cycle (line 6), all data is in the R

```
line procedure BPC(R, B, k);
     {Permute R according to the BPC permutation B}
     {k is the hypercube dimension}

     {Find the cycles of B}
  1  for b := 0 to k - 1 do
  2    if B_b = -b then R(i^(b)) <- R(i)
  3    else if | B_b | ≠ b then
  4    begin
  5    {Follow a nontrivial cycle of B}
  6      j := b; s := B_b;
  7      repeat
  8        q := | B_j |; {Next route is along dimension q}
  9        {Put element to be routed in S}
 10        if B_j ≥ 0 then S(i) :=: R(i), (i_j ≠ i_q)
 11                   else S(i) :=: R(i), (i_j = i_q);
 12        S(i^(q)) <- S(i);
 13        B_j := j;
 14        j := q; {Next position in current cycle of B}
 15      until j = b; {b is start of cycle}

 16      {Move everything back to R}
 17      q := | s |; {Initial B_b value}
 18      if s ≥ 0 then R(i) := S(i), (i_b ≠ i_q)
 19               else R(i) := S(i), (i_b = i_k);
 20    end;
 21  end; {of BPC}
```

Program 2.22 BPC permutations

registers and $OS(i) = DS(i) = *$, $0 \le i < n$. Furthermore, the expression

$$E(i,l) = [((OR(i))_l = i_l) \text{ and } ((OS(i))_l = \overline{i_l})]$$

is true for all i, $0 \le i < n$ and all bit positions l on the new cycle as no routing has as yet been done along any of the corresponding dimensions and $OS(i) = *$. So, following line 6 $E(i,j)$ is true for all i. Assume that $E(i,j)$ is true at line 8. If $B_j \ge 0$, then for PEs i with $i_j \ne i_q$, we have:

$$(DR(i))_q = (OR(i))_j = i_j = \bar{i}_q$$

and

$$(DS(i))_q = (OS(i))_j = \bar{i}_j = i_q$$

For PEs i with $i_j = i_q$, we have:

$$(DR(i))_q = (OR(i))_j = i_j = i_q$$

and

$$(DS(i))_q = (OS(i))_j = \bar{i}_j = \bar{i}_q$$

So, in both cases, following lines 10 and 11, we have:

$$(DR(i))_{q)} = i_q \text{ and } (DS(i))_q \ne i_q$$

A similar proof shows this for the case $B_j < 0$. To complete the proof, we need to show that $E(i,j)$ is true following line 14 and so true at the start of the next iteration of the **repeat** loop. For this, we note that $(OR(i))_j = i_j$ following line 14 as $R(i)$ was not routed along dimension $j = q$ in this iteration of the **repeat** loop and no routes along this dimension were performed previously. Further, $(OS(i))_j = \bar{i}_j$ as the only route ever performed along dimension $j = q$ exchanges the S register data. □

Theorem 2.10 Lines 18 and 19 correctly move records back to the R registers.

Proof: The first time lines 10 and 11 are executed for any nontrivial cycle, half of the PEs move their data to their S registers. Let us refer to these PEs as the *data transfer* PEs. The transfer PEs have an index i such that $i_b \neq i_{|s|}$ if $s = B_b \geq 0$ and $i_b = i_{|s|}$ otherwise. Following the first execution of line 12, these PEs lose their data and all n of the initial data records are in the remaining $n/2$ PEs. The transfer PEs and the remaining PEs define two sets of PEs. In the case of our example of Figure 2.18, the transfer set is {001, 010, 101, 110} and the remaining set is {000, 011, 100, 111}. In one set all PEs have bits b and $|s|$ equal while in the other set these bits are different for all PEs. In the iterations of the **repeat** loop that exclude the first and last, data cannot transfer from one PE set to the other as none of these iterations involves a route along dimensions b or $|s|$ and a route along any other dimension cannot affect the relationship between i_b and $i_{|s|}$. In the last iteration of the **repeat** loop, data is routed along dimension b. As a result, data moves from one PE set to the other. Consequently, when line 17 is reached the transfer PEs have their data in S while the remaining PEs have it in R. Lines 18 and 19 move data from S to R only in the transfer PEs. □

Before concluding this section, we show that procedure *BPC* (Program 2.22) is optimal in the sense that for every BPC permutation B it performs the fewest possible number of routes. Define $\beta(B)$ as below:

$$\beta(B) = |\{b \mid B_b \neq b\}|$$

It is easy to see that the number of routes performed by procedure *BPC* is $\beta(B)$. The following theorem shows that $\beta(B)$ is a lower bound on the number of routes needed to perform B on a hypercube.

Theorem 2.11 (Nassimi and Sahni 1982) $\beta(B)$ is a lower bound on the number of routes needed to perform the BPC permutation B on a hypercube.

Proof: For every b such that $B_b \neq b$, there is at least one i, $0 \leq i < n$ for which $(OR(i))_b \neq (DR(i))_b$. Hence B cannot be performed on a hypercube without a route along dimension b. So, $\beta(B)$ is a lower bound on the number of routes needed. □

2.19 Summary

We have studied many hypercube algorithms in this Chapter. In this section, we provide a summary of these together with where these are used in this book. In the following, M and W are powers of 2. They represent the size (i.e., the number of processors) in a subhypercube. Unless otherwise stated, the size of the full hypercube is denoted by P.

Program 2.1 *Broadcast* (A, d)
Task: Broadcast the data in register A of PE 0 to the A registers of the remaining processors of a dimension d hypercube.
Complexity: $O(d)$.

Program 2.2 *Broadcast* (A, d, k)
Task: Broadcast the data in register A of PE k to the A registers of the remaining processors of a dimension d hypercube. This procedure assumes the availability of a special value *null* that is not the value in the A register of any of the PEs in the hypercube.
Complexity: $O(d)$.

Program 2.3 *WindowBroadcast* (A, k)
Task: An arbitrary dimension hypercube is assumed to be partitioned into subhypercubes of dimension k. The processor indices in each such subhypercube differ only in their least significant k bits. Each subhypercube initially contains data in the A register of its single PE with least significant k bits equal to $m_0, m_1, \ldots, m_{k-1}$. Data from this PE of each subhypercube is broadcast to the remaining PEs of the subhypercube.
Complexity: $O(k)$.
Applications: Matrix multiplication (Programs 3.2, 3.3, and 3.8), template matching (Programs 5.1 and 5.2), clustering (Programs 7.3 through 7.6), and string editing (Programs 9.1, 9.2, and 9.3).

Program 2.4 *WindowBroadcast* (A, k)
Task: Same as that of Program 2.3. The originating PE of the data to be broadcast in each subhypercube can, however, be different. Each PE has a register *selected*. The unique originating PE in each subhypercube has *selected* equal to **true**. In the remaining PEs, this register has the value

false. The procedure assumes the availability of a special value *null* that is not the value in the A register of any of the PEs in the hypercube.

Program 2.5 *SIMDDataSum* (A, k)
Task: The sum of the A register values in each subhypercube of dimension k is computed and stored in the A registers of the PE with the least significant k bits equal to zero in the respective subhypercubes. Despite the procedure's name, it works equally well for both SIMD and MIMD hypercubes.
Complexity: $O(k)$.
Applications: Template matching (Program 5.2), and clustering (Programs 7.5 and 7.7).

Program 2.6 *SIMDAllSum* (A, k)
Task: Same as that of Program 2.5 except that the sum of the A registers of the PEs in each subhypercube is left in the A registers of all PEs in the subhypercube.
Complexity: $O(k)$.
Application: Matrix multiplication (Programs 3.2, 3.3, and 3.8).

Program 2.7 *SIMDPrefixSum* (A, k, S)
Task: This works on each dimension k, $k = \log_2 W$, subhypercube of an SIMD hypercube. The hypercube PE $l = iW + q$, $0 \le q < W$ is the q'th PE in the i'th dimension k subhypercube. This PE computes, in its S register, the sum of the A register values of the 0'th through q'th PEs in its subhypercube, $0 \le q < W$, $0 \le i < w$, where w is the number of subhypercubes of dimension k.
Complexity: $O(k)$.
Note: An algorithm with the same complexity can be written for MIMD hypercubes using a gray code indexing scheme within each subhypercube.
Application: String editing (Program 9.2).

Program 2.8 *SIMDShift* (A, i, W)
Task: Shift the data in the A registers of the PEs of an SIMD hypercube counterclockwise by i PEs. The shift is done independently in each window/subhypercube of size W. The linear ordering of PEs in each subwindow is as per the chain mapping discussed in Chapter 1.
Complexity: $O(\log W)$ for general i and $O(\log(W/i))$ for i a power of 2.
Note: For MIMD hypercubes a shift of i can be done in $O(\log W)$ time for general i and in $O(1)$ time for i a power of 2. For small i it is quicker to perform i shifts of one each. This would take $O(i)$ time. The gray code indexing scheme is used within each subhypercube. If all even length, all odd length, or all possible shifts are to be performed on an SIMD hypercube, these can be done in linear time using a different technique (Section 2.7). A linear time algorithm to perform these shifts on an MIMD hypercube using a gray code mapping results from the repeated use of length two or length one shifts.
Applications: One dimensional convolution (Programs 4.1, 4.4, and 4.5), template matching (Program 5.2), image shrinking (Programs 8.1 and 8.2), image translation (Section 8.3), image rotation (Section 8.4), and string editing (Program 9.1). The linear time odd/even shifts are used in Program 4.6 (O(1) memory SIMD one dimensional convolution).

Program 2.9 *circulate* (A)
Task: Circulate the A register data of the PEs in an SIMD hypercube through all the PEs. The procedure is easily modified to circulate in subhypercubes.
Complexity: $O(P)$ where P is the number of processors in the hypercube or subhypercube.
Note: Data circulation in an MIMD machine can be done in the same time using a much simpler algorithm. This algorithm repeatedly shifts by one and uses the gray code ordering of Chapter 1.
Applications: Consecutive sum (Program 2.10), adjacent sum (Program 2.11), data accumulation (Program 2.12), matrix multiplication (Programs 3.5, 3.6 and 3.8), one dimensional convolution (Programs 4.1 and 4.2), template matching (Program 5.1), and clustering (Program 7.2).

Program 2.10 *SIMDConsecutiveSum* (X, S, M)
Task: Each PE of the SIMD hypercube contains an array of values $X[0..M-1]$. The j'th PE in each window of size M computes the sum of the $X[j]$ values in its window, $0 \le j < M$. The sum computed by each PE is stored in its S register.
Complexity: $O(M)$.
Note: This task can be performed on an MIMD hypercube in the same time using the gray code scheme and repeated shifts of one.
Application: Clustering (Programs 7.3 and 7.4).

Program 2.11 *SIMDAdjacentSum* (A, T, M)
Task: Each PE of the SIMD hypercube has an array $A[0..M-1]$. PE j computes, in its T register, the sum

$$\sum_{i=0}^{M-1} A[i]((p+i) \bmod P)$$

Complexity: $O(M + \log(P/M))$.
Note: This task is easily performed on an MIMD hypercube in $O(M)$ time using the gray code mapping.
Application: Template matching (Program 5.1).

Program 2.12 *SIMDAccum* (A, I, M)
Task: Each PE, j, of the SIMD hypercube accumulates an array A of I values such that

$$A[i](j) = I((j+i) \bmod P),\ 0 \le i < M$$

Complexity: $O(M + \log(P/M))$.
Note: This task is easily performed on an MIMD hypercube in $O(M)$ time using the gray code mapping.
Applications: One dimensional convolution (Programs 4.1, 4.2, and 4.3), and template matching (Programs 5.1 and 5.2).

Program 2.14 *rank* (k)
Task: Rank the selected processors in each size 2^k window of the SIMD hypercube.
Complexity: $O(k)$.

Note: Can also be done in O(k) time on an MIMD hypercube using the gray code order.
Applications: Random access reads (Program 2.20) and writes (Program 2.21).

Program 2.15 *concentrate* (G, k)
Task: Let $G(i).R$ be the rank of each selected processor i in the window of size 2^k that it is contained in. For each selected PE, i, the record $G(i)$ is sent to the $G(i).R$'th PE in the size 2^k window that contains PE i. The procedure needs to be modified for MIMD hypercubes using a gray code scheme.
Complexity: O(k).
Applications: Random access reads (Program 2.20) and writes (Program 2.21).

Program 2.16 *distribute* (G, k)
Task: This is the inverse of a concentrate.
Complexity: O(k).
Applications: Random access reads (Program 2.20) and writes (Program 2.21).

Program 2.17 *generalize* (G, k)
Each record $G(i)$ has a high destination $G(i).H$. The high destinations in each size 2^k window are in ascending order. Assume that $G(-1).H = 0$. The record initially in processor i is routed to processors $G(i-1).H$ through $G(i).H$ of the window provided that $G(i).H \neq \infty$. If $G(i).H = \infty$, then the record is ignored. The procedure as written assumes a PE ordering that corresponds to that generally used for SIMD hypercubes. The procedure needs to be modified for MIMD hypercubes using a gray code ordering.
Complexity: O(k).
Application: Random access reads (Program 2.20).

Program 2.20 Random Access Read
Task: Each PE in an N processor hypercube reads the A register data of some other PE in the hypercube.
Complexity: $O(\log^2 N)$.

Program 2.21 Random Access Write
Task: Each PE in an N processor hypercube sends its A register data to the A register of some other PE in the hypercube.
Complexity: $O(\log^2 N)$ in case of an arbitrary random access write (RAW) or highest/lowest RAW.
Applications: Clustering (Program 7.7), image rotation (Section 8.4), and scaling (Section 8.5).

Program 2.22 $BPC(R, B, k)$
Task: The R register data of the PEs in a k dimension hypercube are permuted according to the BPC permutation B.
Complexity: $O(k)$. Actually, the algorithm is optimal in that for each BPC permutation, B, it uses the fewest possible number of interprocessor routes.
Applications: Clustering (Chapter 7), image transformations (Section 8.1), and string editing (Program 9.3).

Chapter 3

SIMD Matrix Multiplication

3.1 n^3 Processors

Suppose that two $n \times n$ matrices $A[0:n-1, 0:n-1]$ and $B[0:n-1, 0:n-1]$ are to be multiplied on an SIMD hypercube to get the product matrix C where

$$C[i, j] = \sum_{k=0}^{n-1} A[i, k] * B[k, j], \quad 0 \le i, j < n$$

Throughout this chapter we shall assume that n is a power of 2. In this section we assume that the matrix product is to be computed on an SIMD hypercube that has $n^3 = 2^{3q}$ processors. I.e., the hypercube dimension is $3q$. The n^3 processors may be viewed as forming an $n \times n \times n$ array. Hence we have a processor at each of the array positions (k, i, j), $0 \le i, j, k < n$. We shall use two different notations to reference the hypercube processors. One is the usual one dimensional notation. In this PE(l) refers to the l'th hypercube processor, $0 \le l < n^3$. The second is a three dimensional notation. In this PE(k, i, j) refers to the processor in position (k, i, j) of our three dimensional view. The mapping between the one and three dimensional notations is done using row major order (Horowitz and Sahni 1987). If PE(l) and PE(k, i, j) refer to the same PE then l =

$kn^2 + i * n + j$. Furthermore if the binary representation of l is $l_{3q-1} l_{3q-2} \cdots l_0$ then the binary representation of k is $l_{3q-1} l_{3q-2} \cdots l_{2q}$; that of i is $l_{2q-1} l_{2q-1} \cdots l_q$; and that of j is $l_{q-1} l_{q-2} \cdots l_0$. $Q(k, i, j)$ refers to memory location Q of PE(k, i, j) and k, i, and j, respectively, represent dimensions 1, 2, and 3 of the three dimensional view.

We begin with $A(0, i, j) = A[i, j]$ and $B(0, i, j) = B[i, j], 0 \le i, j < n$. The product matrix C is to be stored such that $C(0, i, j) = C[i, j]$, $0 \le i, j < n$. The strategy used in the matrix multiplication algorithm of Dekel, Nassimi, and Sahni (1981) is given in Program 3.1. The objective is to use PE(k, i, j) to compute the product $A[i, k] * B[k, j]$ and then to sum these products over all processors with the same i and j values (i.e, $(*, i, j)$) to get the product values $C[i, j]$. For this the A's and B's need to be distributed as in step 1. Once this has been done the A's and B's are multiplied (Step 2) and the products summed (Step 3).

Step 1: [Distribute data] Distribute the data so that $A(k, i, j) = A[i, k]$ and $B(k, i, j) = B[k, j], 0 \le i, j, k < n$.

Step 2: [Multiply] PE(k, i, j) computes $C(k, i, j) = A(k, i, j) * B(k, i, j) = A[i, k] * B[k, j], 0 \le i, j, k < n$.

Step 3: [Add terms] PE(0, i, j) computes
$$\sum_{k=0}^{n-1} C(k, i, j) = \sum_{k=0}^{n-1} A[i, k] * B[k, j], 0 \le i, j < n.$$

Program 3.1 Steps in the n^3 processor matrix multiplication algorithm of (Dekel, Nassimi, and Sahni 1981)

Procedure *MultCube* (Program 3.2 and Program 3.3) implements this strategy. Program 3.2 uses three dimensional notation while Program 3.3 uses one dimensional notation. Consider Program 3.2. The first **for** loop broadcasts the A's and B's in windows of size n (data is broadcast along the first dimension of the three dimensional interpretation). Following this broadcast we have

$$A(k, i, j) = A[i, j], B(k, i, j) = B[i, j], 0 \le k < n$$

As a result, $A(k, i, k) = A[i, k]$ and $B(k, k, j) = B[k, j]$. To get the desired distribution of Step 1 we, therefore, need to replicate $A(k, i, k)$ over the third dimension and replicate $B(k, k, j)$ over the second dimension. Both these replications are accomplished by data broadcasts in appropriate windows. The second **for** loop of procedure *MultCube* does this for A and the third one does this for B. Step 2 is a simple multiplication. Step 3 is implemented as a data sum in windows of size n. This is done by the fourth **for** loop. The complexity of the procedure is readily seen to be $O(q) = O(\log n)$.

3.2 n^2 Processors

When $n^2 = 2^{2q}$ processors are available two $n \times n$ matrices can be multiplied in $O(n)$ time (Dekel, Nassimi, and Sahni 1981). As in the previous section, we shall use two notations: a one dimensional notation and a two dimensional notation. In the two dimensional notation we view the hypercube as an $n \times n$ array of processors. This is done using the standard row major mapping between one and two dimesional arrays (Horowitz and Sahni 1987). In this mapping, PE(l) and PE(i, j) are the same physical PE iff $l = i * n + j$.

Initially, $A(i, j) = A[i, j]$ and $B(i, j) = B[i, j], 0 \le i, j < n$. The final condition is $C(i, j) = C[i, j], 0 \le i, j < n$. The matrix multiplication algorithm of Dekel, Nassimi, and Sahni (1981) consists of two steps (Program 3.4). The overall strategy is to have PE(i, j) compute all n terms in the sum for $C[i, j], 0 \le i, j < n$. Since, initially, the A and B locations of PE(i, j) are such that their product is not one of the terms in the sum for $C[i, j]$, the first step of the algorithm of Dekel, Nassimi, and Sahni (1981) aligns the A's and B's so that following the alignment each processor contains an A and a B value whose product is one of the terms in the C value to be computed by that processor. The particular alignment scheme used results in

$$A(i, j) = A[i, i \oplus j], B(i, j) = B[i \oplus j, j], 0 \le i, j < n$$

---

procedure *MultCube* (A, B, C);
{Multiply the $n \times n$ matrices A and B on an n^3 processor SIMD hypercube}
{Three dimensional notation}
begin
  {Step 1: Distribute A and B}
  {Broadcast $A(0, i, j)$ and $B(0, i, j)$ on dimension 1}
  for $m := 0$ **to** $q - 1$ **do**
  begin
    $A(k^{(m)}, i, j) \leftarrow A(k, i, j), \ (k_m = 0)$;
    $B(k^{(m)}, i, j) \leftarrow B(k, i, j), \ (k_m = 0)$;
  end;

  {Broadcast $A(k, i, k)$ on dimension 3}
  for $m := 0$ **to** $q - 1$ **do**
    $A(k, i, j^{(m)}) \leftarrow A(k, i, j), \ (k_m = j_m)$;

  {Broadcast $B(k, k, j)$ on dimension 2}
  for $m := 0$ **to** $q - 1$ **do**
    $B(k, i^{(m)}, j) \leftarrow B(k, i, j), \ (k_m = i_m)$;

  {Step 2: Multiply}
  $C(k, i, j) := A(k, i, j) * B(k, i, j)$;

  {Step 3: Sum C's on dimension 1}
  for $m := 0$ **to** $q - 1$ **do**
    $C(k, i, j) \leftarrow C(k, i, j) + C(k^{(m)}, i, j)$;
end; {of *MultCube*}

Program 3.2 Matrix multiplication with n^3 processors (three dimensional notation)

```
procedure MultCube (A, B, C);
{Multiply the n × n matrices A and B on an n^3 processor SIMD hypercube}
{One dimensional notation}
begin
  {Step 1: Distribute A and B}
  {Broadcast A(0, i, j) and B(0, i, j) on dimension 1}
  for m := 2q to 3q - 1 do
  begin
    A(a^(m)) ← A(a), (a_m = 0);
    B(a^(m)) ← B(a), (a_m = 0);
  end;

  {Broadcast A(k, i, k) on dimension 3}
  for m := 0 to q - 1 do
    A(a^(m)) ← A(a), (a_m = a_{2q+m});

  {Broadcast B(k, k, j) on dimension 2}
  for m := q to 2q - 1 do
    B(a^(m)) ← B(a), (a_m = a_{q+m});

  {Step 2: Multiply}
  C(a) := A(a) * B(a);

  {Step 3: Sum C's on dimension 1}
  for m := 2q to 3q - 1 do
    C(a) ← C(a) + C(a^(m));
end; {of MultCube}
```

Program 3.3 Matrix multiplication with n^3 processors (one dimensional notation)

where $\oplus$ is the exclusive or operator.

Step 1: [Alignment] Align the A's and B's so that PE(i, j) contains an A and a B value whose product is one of the terms in the sum for $C[i, j], 0 \leq i, j < n$.

Step 2: [Shift-multiply-add]
Initialize $C(i, j)$ to $A(i, j) * B(i, j)$.
for $m := 1$ **to** $n - 1$ **do**
begin
  Move the A's and B's so that each processor has an A and a B whose product is a new term in the sum for $C[i, j]$.
  Multiply the A and B value in each processor and add to the C value.
end;

Program 3.4 Steps in the n^2 processor matrix multiplication algorithm of Dekel, Nassimi, and Sahni (1981)

As an example consider the case $n = 4$. Figure 3.1 (a) shows the initial configuration. The 16 processor hypercube is shown as a 4×4 array of processors. Each processor is represented as a square. The first two bits of a processor index are shown at the left of each processor row and the last two (i.e., least significant) at the top of each column. In the two dimensional notation the first two bits give the i index while the last two give the j index. Within each processor the index of the A and B values it contains is given. This is given in decimal notation. So, for example, PE(10, 11) = PE(2, 3) contains $A[2, 3]$ and $B[2, 3]$ initially. The configuration following the alignment step is given in Figure 3.2 (b). From the definition of the alignment pattern, PE(10, 11) is to contain $A[10, 10 \oplus 11] = A[10, 01] = A[2, 1]$ and $B[10 \oplus 11, 11] = B[01, 11] = B[1, 3]$.

	00	01	10	11	
00	00	01	02	03	A
	00	01	02	03	B
01	10	11	12	13	A
	10	11	12	13	B
10	20	21	22	23	A
	20	21	22	23	B
11	30	31	32	33	A
	30	31	32	33	B

(a) Initial

	00	01	10	11
00	00	01	02	03
	00	11	22	33
01	11	10	13	12
	10	01	32	23
10	22	23	20	21
	20	31	02	13
11	33	32	31	30
	30	21	12	03

(b) After alignment

01	00	03	02
10	01	32	23
10	11	12	13
00	11	22	33
23	22	21	20
30	21	12	03
32	33	30	31
20	31	02	13

(c) Bit 0 exchange

03	02	01	00
30	21	12	03
12	13	10	11
20	31	02	13
21	20	23	22
10	01	32	23
30	31	32	33
00	11	22	33

(d) Bit 1 exchange

Figure 3.1 n^2 processor matrix multiplication (continued)

The A and B moving of Step 2 is done using the data circulation scheme of Section 2.6. The A values are circulated in rows and the B values in columns. This ensures that processors in row i always have an A value from row i of matrix A and processors in column j always have a B value from column j of matrix B. Since each row and column is a hypercube of dimension two, the function $f(2, *) = 0, 1, 0$ is used. First the A values are exchanged along bit 0 of the row index and the B values

02 20	03 31	00 02	01 13
13 30	12 21	11 12	10 03
20 00	21 11	22 22	23 33
31 10	30 01	33 32	32 23

(e) Bit 0 exchange

Figure 3.1 n^2 processor matrix multiplication

are exchanged along bit 0 of the column index. The result is shown in Figure 3.1 (c). The next exchange is along bit 1 (Figure 3.1 (d)) and the final exchange is along bit 0 (Figure 3.1 (e)). As can be seen each processor (i, j) is able to compute a new product term in the sum for $C[i, j]$ following each exchange. Hence following the last exchange each processor can complete the computation of the C matrix entry it was assigned to compute.

Procedure *MultSquare* (Program 3.5 and Program 3.6) implements the strategy of Program 3.4. Program 3.5 uses two dimensional notation while Program 3.6 uses one dimensional notation.

To establish the correctness of Program 3.5 we need to show that following the first **for** loop the data has been aligned as stated above and that at the end of each iteration of the second **for** loop each PE has an A and a B value whose product is a term in the sum it is to compute. The fact that PE(i, j) computes different terms in each iteration follows from the correctness of the data circulation algorithm of Section 2.6.

```
procedure MultSquare (A, B, C);
{Matrix multiplication with n^2 processors}
{Two dimensional notation}
begin
   {Step 1: Align data}
   for m := 0 to q - 1 do
   begin
      A (i, j^(m)) <- A (i, j), (i_m = 1);
      B (i^(m), j) <- B (i, j), (j_m = 1);
   end;

   {Step 2: Shift-multiply-add}
   C (i, j) := A (i, j) * B (i, j);
   for m := 1 to n - 1 do
   begin
      l := f (q, m); {circulation function}
      A (i, j^(l)) <- A (i, j);
      B (i^(l), j) <- B (i, j);
      C (i, j) := C (i, j) + A (i, j) * B (i, j);
   end;
end; {of MultSquare}
```

Program 3.5 Matrix multiplication with n^2 processors (two dimensional notation)

Lemma 3.1 (Dekel, Nassimi, and Sahni 1981) Following the first **for** loop of Program 3.5 we have

$$A(i, j) = A[i, i \oplus j], B(i, j) = B[i \oplus j, j], 0 \le i, j < n$$

Proof: We provide the proof only for A. The proof for B is similar. Let $d = i \oplus j$. The A value initially in PE(i, j) is to be routed to PE(i, d). From the definition of the exclusive or operator we obtain $d_m = j_m$ if $i_m = 0$ and $d_m = \bar{j}_m$ if $i_m = 1$. So when $i_m = 0$ no routing along bit m of the column index is needed. When $i_m = 1$ we need to route along bit m. The first **for** loop of Program 3.5 does precisely this. □

```
procedure MultSquare (A, B, C);
{Matrix multiplication with n^2 processors}
{One dimensional notation}
begin
  {Step 1: Align data}
  for m := 0 to q - 1 do
  begin
    k := q + m;
    A (a^(m)) ← A (a), (a_k = 1);
    B (a^(k)) ← B (a), (a_m = 1);
  end;

  {Step 2: Shift-multiply-add}
  C (a) := A (a) * B (a);
  for m := 1 to n - 1 do
  begin
    l := f (q, m); {circulation function}
    k := q + l;
    A (a^(l)) ← A (a);
    B (a^(k)) ← B (a);
    C (a) := C (a) + A (a) * B (a);
  end;
end; {of MultSquare}
```

Program 3.6 Matrix multiplication with n^2 processors (one dimensional notation)

Lemma 3.2 (Dekel, Nassimi, and Sahni 1981) At the end of each iteration of the second **for** loop of Program 3.5 each PE has an A and a B value whose product is a term in the sum it is to compute.

Proof: Consider any processor (i, j). Let $index(j, m)$ and $index(i, m)$ be such that $A[i, index(j, m)]$ and $B[index(i, m), j]$ are in PE(i, j) following iteration m of the second **for** loop. For $m = 0$ we have $index(j, 0) = index(i, 0) = i \oplus j$. Since the same sequence of moves is made on the rows of A as on the columns of B it follows from the proof of Theorem 2.1 that $index(j, m) = index(i, m)$. So after iteration m the A and B values in PE(i, j) are indeed such that their product is one of the terms in the sum for $C[i, j]$. □

3.3 n^2r, $1 \leq r \leq n$ Processors

The strategies used in procedures *MultCube* and *MultSquare* can be combined to arrive at an efficient matrix multiplication algorithm for the case when $n^2r = 2^{2q+s}$ ($n = 2^q$ and $r = 2^s$) processors are available. The complexity of the combined algorithm is $O(n/r + \log r)$. Suppose we partition the matrices A, B, and C into blocks of size $n/r \times n/r$ each. This is done in a natural way by tiling the matrices using an $n/r \times n/r$ window. The partitioning results in r^2 partitions $PX[i, j]$, $0 \leq i, j < r$, where $X \in \{A, B, C\}$. Figure 3.3 shows the partitions of A superimposed over the original $n \times n$ matrix A. The figure is for the case $r = 4$. The partition $PC[i, j]$ of the product matrix C is given by the formula

$$PC[i, j] = \sum_{k=0}^{r-1} PA[i, k] * PB[k, j]$$

where $PA[i, k] * PB[k, j]$ is the product of two $n/r \times n/r$ matrices.

The n^2r processors of the hypercube may be viewed as an $r \times r \times r$ array of superprocessors SP(k, i, j), $0 \leq i, j, k < r$, where each superprocessor represents $(n/r)^2$ normal hypercube processors. The mapping from a normal processor PE(a), $0 \leq a < n^2r$ to a superprocessor and within a superprocessor to an individual processor is done in the following way.

PA_{00}	PA_{01}	PA_{02}	PA_{03}
PA_{10}	PA_{11}	PA_{12}	PA_{13}
PA_{20}	PA_{21}	PA_{22}	PA_{23}
PA_{30}	PA_{31}	PA_{32}	PA_{33}

A

Figure 3.3 A 4×4 partitioning of matrix A

Let $a_{2q+s-1}a_{2q+s-2} \cdots a_0$ be the binary representation of a. PE(a) is a processor of the superprocessor SP(k, i, j) iff $a_{2q+s-1}a_{2q+s-2} \cdots a_{2q}$, $a_{2q-1}a_{2q-2} \cdots a_{2q-s}$, and $a_{q-1}a_{q-2} \cdots a_{q-s}$ are, respectively, the binary representations of k, i, and j. Furthermore $a_{2q-s-1}a_{2q-s-2} \cdots a_q$ and $a_{q-s-1}a_{q-s-2} \cdots a_0$ are, respectively, the row and column number of the individual processor within the superprocessor.

We begin with PA[i, j] and PB[i, j] stored in superprocessor SP[0, i, j], $0 \le i, j < r$. Each of the $(n/r)^2$ terms in a matrix partition is stored in a single processor within the superprocessor. The mapping is the natural one with the term in row b and column c of the matrix partition being stored in the processor at row b and column c of the superprocessor. Using the bit decomposition provided above the initial configuration is easy to specify using one dimensional notation. Initially we have

$$A(i * n + j) = A[i, j] \text{ and } B(i * n + j) = B[i, j], 0 \le i, j < n$$

At the top level of the matrix multiplication algorithm we work with superprocessors and matrix partitions. Since the partitions are $r \times r$ matrices and the superprocessors form an $r \times r \times r$ array the steps of our n^3 processor algorithm may be used. These are restated in Program 3.7 using superprocessor and matrix partition terminology.

Step 1: [Distribute matrix partitions] Distribute the partitions so that $PA(k, i, j) = PA[i, k]$ and $PB(k, i, j) = PB[k, j]$, $0 \le i, j, k < r$.

Step 2: [Multiply] SP(k, i, j) computes $PC(k, i, j) = PA(k, i, j) * PB(k, i, j) = PA[i, k] * PB[k, j]$, $0 \le i, j, k < r$.

Step 3: [Add matrix partitions] SP(0, i, j) computes $\sum_{k=0}^{r-1} PC(k, i, j) = \sum_{k=0}^{r-1} PA[i, k] * PB[k, j]$, $0 \le i, j < r$.

Program 3.7 Steps in matrix multiplication algorithm using superprocessors

Step 2 of Figure 3.4 requires us to multiply two matrix partitions stored in a superprocessor. Since a matrix partition is an $n/r \times n/r$ matrix and a superprocessor is also viewed as an $n/r \times n/r$ array of processors this matrix product may be performed using the steps in Program 3.4. Putting these ideas together and keeping the mapping between the one dimensional indexing scheme and the three dimensional one for superprocessors and the two dimensional scheme for processors within a superprocessor we get the procedure of Program 3.8. Its correctness follows from that of Program 3.3 and Program 3.6.

---

procedure *MatrixMultiply* (A, B, C, n, q, r, s);
{Matrix multiplication with n^2r, $1 \le r \le n$ processors}
{$n = 2^q$ and $r = 2^s$}
begin
{Step 1: Distribute PA and PB}
{Broadcast $PA(0, i, j)$ and $PB(0, i, j)$ on dimension 1}
{of the superprocessors}
for $m := 2q$ **to** $2q + s - 1$ **do**
begin
$A(a^{(m)}) \leftarrow A(a), \ (a_m = 0)$;
$B(a^{(m)}) \leftarrow B(a), \ (a_m = 0)$;
end;

{Broadcast $PA(k, i, k)$ on dimension 3 of the superprocessors}
for $m := q - s$ **to** $q - 1$ **do**
$A(a^{(m)}) \leftarrow A(a), \ (a_m = a_{q+s+m})$;

{Broadcast $PB(k, k, j)$ on dimension 2 of the superprocessors}
for $m := 2q - s$ **to** $2q - 1$ **do**
$B(a^{(m)}) \leftarrow B(a), \ (a_m = a_{s+m})$;

{Step 2: Multiply each $n/r \times n/r$ partition}
{Step 2.1: Align data}
for $m := 0$ **to** $q - s - 1$ **do**
begin
$k := q + m$;
$A(a^{(m)}) \leftarrow A(a), \ (a_k = 1)$;
$B(a^{(k)}) \leftarrow B(a), \ (a_m = 1)$;
end;

{Step 2.2: Shift-multiply-add}
$C(a) := A(a) * B(a)$;
for $m := 1$ **to** $n/r - 1$ **do**
begin
$l := f(q-s-1, m)$; {circulation function}

```
      k := q + l;
      A(a^(l)) <- A(a);
      B(a^(k)) <- B(a);
      C(a) := C(a) + A(a) * B(a);
   end;

   {Step 3: Add product partitions along dimension 1{
   {of the superprocessors}
   for m := 2q to 2q + s - 1 do
      C(a) <- C(a) + C(a^(m));
end; {of MatrixMultiply}
```

Program 3.8 Matrix multiplication with n^2r processors

The complexity of procedure *MatrixMultiply* (Program 3.8) is seen to be $O(n/r + \log r)$. Furthermore when $r = n$ this procedure works exactly as Program 3.3 and when $r = 1$, it works exactly as Program 3.6. Interestingly, when $r = n/\log n$ the complexity of procedure *MatrixMultiply* is $O(\log n)$ and the processor-time product is n^3 which is the same as that for the single processor algorithm on which it is based.

3.4 r^2, $1 \le r < n$ Processors

This case is efficiently handled by partitioning A, B, and C into r^2 partitions each of which is an $n/r \times n/r$ matrix. $PX[i, j]$, $X \in \{A, B, C\}$, is stored in PE(i, j). I.e., each PE stores $(n/r)^2$ elements of each of the matrices A, B, and C. With this partitioning and storage scheme we have an $r \times r$ array of matrix partitions and an $r \times r$ array of hypercube processors with each processor holding the corresponding partitions of A, B, and C. The situation is identical to that of Section 3.2 except that processors hold partitions rather than single elements.

Matrix multiplication proceeds exactly as in procedure *MultSquare* except that in each route a matrix partition rather than a single matrix element is routed. Also, whenever procedure *MultSquare* multiplies two elements of A and B and adds to C, the new algorithm will need to multiply a matrix partition of A with one of B and add to a matrix partition of C. We leave the development of the formal procedure as an exercise. If $t(n)$ is the time needed to multiply two $n \times n$ matrices using a single processor then the resulting hypercube algorithm for the case of r^2 processors, $1 \le r < n$ has complexity $O(n^2/r + rt(n/r))$. If the matrix partitions are multiplied using the classical single processor algorithm, then $t(n) = O(n^3)$ and the complexity of our r^2 processor multiplication algorithm becomes $O(n^2/r + n^3/r^2) = O(n^3/r^2)$.

3.5 Summary

The performance characteristics of the four matrix multiplication algorithms are summarized in Figure 3.5. The speedup and efficiency are computed relative to the classical uniprocessor matrix multiplication algorithm whose complexity is $O(n^3)$. It is interesting to note that the efficiency of the n^2 and r^2 processor algorithms is $O(1)$. The efficiency of the n^2r processor algorithm is also $O(1)$ so long as $r\log r \le n$. While the complexity of this algorithm remains $O(\log n)$ for r in the range from $n/\log n$ to n, its efficiency declines from $O(1)$ to $O(1/\log n)$.

#Processors	Complexity	Speedup	Efficiency
n^3	$O(\log n)$	$O(n^3/\log n)$	$O(1/\log n)$
n^2	$O(n)$	$O(n^2)$	$O(1)$
$n^2r,\ 1 \le r \le n$	$O(n/r + \log r)$	$O(n^3/(n/r + \log r))$	$O(n/(n + r\log r))$
$r^2,\ 1 \le r < n$	$O(n^3/r^2)$	$O(r^2)$	$O(1)$

Figure 3.5 Performance of the matrix multiplication algorithms

Chapter 4

One Dimensional Convolution

4.1 The Problem

The inputs to the one dimensional convolution problem are vectors $I[0..N-1]$ and $T[0..M-1]$. The output is the vector $C1D$ where

$$C1D[i] = \sum_{v=0}^{M-1} I[(i+v) \bmod N] * T[v],\ 0 \le i < N$$

In this chapter we develop algorithms to compute $C1D$ on SIMD and MIMD hypercubes that have N processors. We assume that M is a power of 2 and consider the following two cases for the amount of memory available on each hypercube processor

(1) Each PE has $O(M)$ memory

(2) Each PE has $O(1)$ memory

Our algorithms assume that the vector I is mapped onto the hypercube using the identity mapping $I[i]$ on PE(i) in the case of an SIMD hypercube and using the gray code mapping $I[i]$ on PE $gray(i)$ (Section 1.2.2) for MIMD hypercubes.

All but one of our algorithms assume that there are initially (N/M) copies of T in the hypercube with one copy in each block/window of M processors. Within a block, the mapping of T is the same as that of I. Figure 4.1 shows the initial data distribution for the case $N = 16$ and $M = 4$. Lines 1, 3, and 4 define the distribution for an SIMD hypercube while lines 2, 3, and 4 do this for an MIMD hypercube. p ($gray(p)$) is the processor index for the case of an SIMD (MIMD) hypercube. The remaining algorithm assumes that T is initially in the control unit of an SIMD hypercube and the control unit is used to broadcast the T values to the hypercube processors as needed. This broadcast is done using an available O(1) time control unit to processors data broadcast feature.

p	0	1	2	3	4	5	6	7	8	9	10	11	12	13	15	16
$gray(p)$	0	1	3	2	6	7	5	4	12	13	15	14	10	11	9	8
I	I_0	I_1	I_2	I_3	I_4	I_5	I_6	I_7	I_8	I_9	I_{10}	I_{11}	I_{12}	I_{13}	I_{14}	I_{15}
T	T_0	T_1	T_2	T_3	T_0	T_1	T_2	T_3	T_0	T_1	T_2	T_3	T_0	T_1	T_2	T_3

$$I_q = I[q]$$

Figure 4.1 Initial data distribution for $N = 16$ and $M = 4$

4.2 O(M) Memory Algorithms

When each processor has $O(M)$ memory, the most effective way to compute $C1D$ is to first perform a data accumulation on I. Following this, each processor has all the I values needed to compute the corresponding entry of $C1D$. Next, the T values are circulated through each block of M processors. During this circulation, the T values are multiplied by I values and the $C1D$ values computed. Procedure *MIMD_C1D_M* (Program 4.1), (Ranka and Sahni 1988a), provides the details for the case of an MIMD hypercube while *SIMD_C1D_M* (Program 4.2), (Ranka and Sahni 1988b), is for an SIMD hypercube.

```
procedure MIMD_C1D_M(M);
{MIMD O(M) memory algorithm for one dimensional convolution}
begin
  MIMDAccum(A, I, M);
  b := igray(p) mod M; {relative index of PE in M block}
  C1D(p) := 0;
  for j := 1 to M do
  begin
    C1D(p) := C1D(p) + A[b](p) * T(p);
    MIMDShift(T, -1, M);
    b := (b+1) mod M;
  end;
end; {of MIMD_C1D_M}
```

Program 4.1 $O(M)$ memory MIMD computation of $C1D$

In procedure *MIMD_C1D_M* the call to *MIMDAccum* accumulates in array $A[*](p)$ the M image values that processor p needs to compute $C1D(p)$. The function *igray* computes the inverse of the gray code mapping. For example, if $gray(8) = 12$, then $igray(12) = 8$. In the **for** loop the $C1D$ value is computed by circulating the T's in windows of size M. Recall that in the case of an MIMD hypercube data circulation is done by repeatedly shifting the data by 1. The time complexity of *MIMD_C1D_M*

---

procedure *SIMD_C1D_M* (*M*);
{SIMD O(*M*) memory one dimensional convolution}
begin
  SIMDAccum (*A*, *I*, *M*);
  $b := p$ **mod** M; {b = index of T in processor p}
  $C1D(p) := 0$;
  for $j := 1$ **to** M **do**
  begin
    $C1D(p) := CID(p) + A[b](p) * T(p)$;
    $l := f(\log_2 M, j)$;
    $T(p^{(l)}) \leftarrow T(p)$;
    $b := b \oplus 2^l$;
  end;
end; { of *SIMD_C1D_M*}

Program 4.2 O(M) memory SIMD computation of $C1D$

is O(M).

Procedure *SIMD_C1D_M* is quite similar to its MIMD counterpart. The essential differences are that it uses the SIMD data accumulation and data circulation algorithms. The variable b in *SIMD_C1D_M* gives the index of the PE from which the current T value originated and f is the data circulation function of Section 2.6. The data accumulation takes O($M + \log(N/M)$) time and the **for** loop takes O(M) time. The overall complexity is therefore O($M + \log(N/M)$).

We can avoid the circulation of T's on an SIMD hypercube if the T vector is initially in the control unit. After the accumulation of the I values in the A arrays of the processors, the control unit can broadcast the T values one at a time to all hypercube processors. When a hypercube processor receives a T value it multiplies it with the appropriate A value and adds the result to its partially computed $C1D$ value. The formal procedure (Prasannan Kumar and Krishnan 1987) is given in Program 4.3.

Making the assumption that a broadcast from the control unit to all the hypercube processors takes O(1) time, the complexity of Program 4.3 is $O(M + \log(N/M))$.

```
procedure Broadcast_C1D_M(M);
{SIMD O(M) memory one dimensional convolution}
{Uses control unit to processors data broadcast feature}
begin
  SIMDAccum(A, I, M);
  C1D(p) := 0;
  for j := 0 to M - 1 do
  begin
    broadcast(T[j]); {control unit broadcasts T[j]}
    {Let T be the value just received from the control unit}
    C1D(p) := CID(p) + A[j](p) * T;
  end;
end; { of Broadcast_C1D_M}
```

Program 4.3 O(*M*) memory SIMD computation of *C*1*D* using data broadcast

4.3 O(1) Memory MIMD Algorithm

When only O(1) memory per PE is available, we begin by first pairing I values in the processors. The pair in processor p is $(A(p), B(p)) = (I[(jM + 2k) \bmod N], I[(jM + 2k + 1) \bmod N])$ where $i = igray(p)$, $j = \lfloor i/M \rfloor$, and $k = i \bmod M$. Figure 4.2 gives the initial *AB* pairs in each PE for the case $N = 16, M = 4$. Notice that following the *AB* pairing, each block of M processors contains all the image values needed to compute its C1D values. The *AB* pairing can be done by first concentrating the even image values in windows of size M and then concentrating the odd values also in windows of size M. When $N = 16$ and $M = 4$, these two data concentration operations result in the configuration of the last column of Figure 4.3. This figure also shows how the two data concentrations can be done

p	$i = igray(p)$	j	k	I	AB
0	0	0	0	I_0	$I_0 I_1$
1	1	0	1	I_1	$I_2 I_3$
3	2	0	2	I_2	$I_4 I_5$
2	3	0	3	I_3	$I_6 I_7$
6	4	1	0	I_4	$I_4 I_5$
7	5	1	1	I_5	$I_6 I_7$
5	6	1	2	I_6	$I_8 I_9$
4	7	1	3	I_7	$I_{10} I_{11}$
12	8	2	0	I_8	$I_8 I_9$
13	9	2	1	I_9	$I_{10} I_{11}$
15	10	2	2	I_{10}	$I_{12} I_{13}$
14	11	2	3	I_{11}	$I_{14} I_{15}$
10	12	3	0	I_{12}	$I_{12} I_{13}$
11	13	3	1	I_{13}	$I_{14} I_{15}$
9	14	3	2	I_{14}	$I_0 I_1$
8	15	3	3	I_{15}	$I_2 I_3$

$I_q = I[q]$

Figure 4.2 Initial *AB* pairs for $N = 16, M = 4$

together. First the *A* and *B* values in each processor are initialized to the current *I* value. Next the *B* values are shifted by –1. Following this, the even processors have the desired *AB* pairs. These are then concentrated in windows of size $M = 4$. A shift of –1 on the *AB* pairs suffices for this. Figure 4.4 shows the steps for the case $N = 16$ and $M = 8$. In this case the concentration is done in windows of size 8. For this the pairs in the even processors are first shifted by –1 and then by –2. If the *AB* concentration is followed by a shift of $-M/2$ on the *AB* registers such that only PEs with the '-' values in the last columns of Figure 4.3 and Figure 4.4 update their *AB* values, the desired pairing of *I* values is obtained. The formal algorithm is given in Program 4.4. Its complexity is $O(\log M)$.

i=*igray* (p)	M block	initial I	shift to B	shift −1	shift −1
				$i_0 = 1$	$i_1 = 1$
		A	B	AB	AB
0	0	I_0	I_0	$I_0 I_1$	$I_0 I_1$
1	0	I_1	I_1	-	$I_2 I_3$
2	0	I_2	I_2	$I_2 I_3$	-
3	0	I_3	I_3	-	-
4	1	I_4	I_4	$I_4 I_5$	$I_4 I_5$
5	1	I_5	I_5	-	$I_6 I_7$
6	1	I_6	I_6	$I_6 I_7$	-
7	1	I_7	I_7	-	-
8	2	I_8	I_8	$I_8 I_9$	$I_8 I_9$
9	2	I_9	I_9	-	$I_{10} I_{11}$
10	2	I_{10}	I_{10}	$I_{10} I_{11}$	-
11	2	I_{11}	I_{11}	-	-
12	3	I_{12}	I_{12}	$I_{12} I_{13}$	$I_{12} I_{13}$
13	3	I_{13}	I_{13}	-	$I_{14} I_{15}$
14	3	I_{14}	I_{14}	$I_{14} I_{15}$	-
15	3	I_{15}	I_{15}	-	-

$I_q = I[q]$

Figure 4.3 Related pairing for $N = 16$, $M = 4$

Once the AB pairing has been done $C1D$ may be computed by rotating the AB values clockwise in a window of size N (in a single rotation, B's move to A's in the same PE and A's move to B's of the next PE) and rotating the T values clockwise in a window of size M. Figure 4.5 shows the initial AB pairs and T values for the case $N = 16$ and $M = 4$. At all times, the product of $A(p)$ and $T(p)$ gives one of the terms needed to compute $C1D(igray(p))$ for every PE p. $B(p)$ will be the next I value needed. Initially, this is true for all processors except those with $igray(p) \bmod M = M - 1$. This situation is remedied by replacing B with I in these processors to get the first column labeled AB'. Following a

i=$igray(p)$	M block	initial I	shift to B	shift -1	shift -1	shift -2
				$i_0 = 1$	$i_1 = 1$	$i_2 = 1$
		A	B	AB	AB	AB
0	0	I_0	I_0	$I_0 I_1$	$I_0 I_1$	$I_0 I_1$
1	0	I_1	I_1	-	$I_2 I_3$	$I_2 I_3$
2	0	I_2	I_2	$I_2 I_3$	-	$I_4 I_5$
3	0	I_3	I_3	-	-	$I_6 I_7$
4	0	I_4	I_4	$I_4 I_5$	$I_4 I_5$	-
5	0	I_5	I_5	-	$I_6 I_7$	-
6	0	I_6	I_6	$I_6 I_7$	-	-
7	0	I_7	I_7	-	-	-
8	1	I_8	I_8	$I_8 I_9$	$I_8 I_9$	$I_8 I_9$
9	1	I_9	I_9	-	$I_{10} I_{11}$	$I_{10} I_{11}$
10	1	I_{10}	I_{10}	$I_{10} I_{11}$	-	$I_{12} I_{13}$
11	1	I_{11}	I_{11}	-	-	$I_{14} I_{15}$
12	1	I_{12}	I_{12}	$I_{12} I_{13}$	$I_{12} I_{13}$	-
13	1	I_{13}	I_{13}	-	$I_{14} I_{15}$	-
14	1	I_{14}	I_{14}	$I_{14} I_{15}$	-	-
15	1	I_{15}	I_{15}	-	-	-

Figure 4.4 Related pairing for $N = 16$, $M = 8$

rotation of AB, we get the second column labeled AB. Now, the B value in processors with $igray(p)$ **mod** $M = M - 2$ needs to be changed to $I(p)$. With this insight, one arrives at procedure *MIMD_C1D_1* (Program 4.5), (Ranka and Sahni 1988a). Its correctness is easily established. The complexity of this procedure is readily seen to be O(M).

```
procedure pairing (M);
{pairing I values in AB registers}
begin
  i := igray (p); {p is processor index}

  {Create needed AB pairs in even processors}
  A (p) := I (p);
  B (p) := I (p);
  MIMDShift (B, −1, P);

  {Concentrate from even processors in windows of size M}
  for j := 1 to logM − 1 do
  begin
    C (p) := B (p); MIMDShift (B, −2^(j−1), M);
    B (p) := C (p), (p_j = 0);
    C (p) := A (p); MIMDShift (A, −2^(j−1), M);
    A (p) := C (p), (p_j = 0);
  end;

  {Shift pairs by −M/2 to get them to remaining processors}
  C (p) := B (p); MIMDShift (B, −(M/2), P);
  B (p) := C (p), (p_(logM − 1) = 0);
  C (p) := A (p); MIMDShift (A, −(M/2), P);
  A (p) := C (p), (p_(logM − 1) = 0);
end; {of pairing}
```

Program 4.4 Pairing of the I's

4.4 O(1) Memory SIMD Algorithm

The startegy here is to have each PE compute two quantities A and B. For any PE, A is the sum of all the $C1D$ terms that are in the M block containing the PE. B is the sum of all $C1D$ terms that are needed by the corresponding PE in the previous M block. The terms contributing to A and B are shown in Figure 4.6. The A and B values are computed in two

i	I	AB	T	AB'	AB	T	AB'	AB	T	AB'	AB	T	AB'
0	I_0	I_0I_1	T_0	I_0I_1	I_1I_2	T_1	I_1I_2	I_2I_3	T_2	I_2I_3	I_3I_4	T_3	I_3I_0
1	I_1	I_2I_3	T_1	I_2I_3	I_3I_4	T_2	I_3I_4	I_4I_5	T_3	I_4I_1	I_1I_2	T_0	I_1I_2
2	I_2	I_4I_5	T_2	I_4I_5	I_5I_6	T_3	I_5I_2	I_2I_3	T_0	I_2I_3	I_3I_4	T_1	I_3I_4
3	I_3	I_6I_7	T_3	I_6I_3	I_3I_4	T_0	I_3I_4	I_4I_5	T_1	I_4I_5	I_5I_6	T_2	I_5I_6
4	I_4	I_4I_5	T_0	I_4I_5	I_5I_6	T_1	I_5I_6	I_6I_7	T_2	I_6I_7	I_7I_8	T_3	I_7I_4
5	I_5	I_6I_7	T_1	I_6I_7	I_7I_8	T_2	I_7I_8	I_8I_9	T_3	I_8I_5	I_5I_6	T_0	I_5I_6
6	I_6	I_8I_9	T_2	I_8I_9	I_9I_{10}	T_3	I_9I_6	I_6I_7	T_0	I_6I_7	I_7I_8	T_1	I_7I_8
7	I_7	$I_{10}I_{11}$	T_3	$I_{10}I_7$	I_7I_8	T_0	I_7I_8	I_8I_9	T_1	I_8I_9	I_9I_{10}	T_2	I_9I_{10}
8	I_8	I_8I_9	T_0	I_8I_9	I_9I_{10}	T_1	I_9I_{10}	$I_{10}I_{11}$	T_2	$I_{10}I_{11}$	$I_{11}I_{12}$	T_3	$I_{11}I_8$
9	I_9	$I_{10}I_{11}$	T_1	$I_{10}I_{11}$	$I_{11}I_{12}$	T_2	$I_{11}I_{12}$	$I_{12}I_{13}$	T_3	$I_{12}I_9$	I_9I_{10}	T_0	I_9I_{10}
10	I_{10}	$I_{12}I_{13}$	T_2	$I_{12}I_{13}$	$I_{13}I_{14}$	T_3	$I_{13}I_{10}$	$I_{10}I_{11}$	T_0	$I_{10}I_{11}$	$I_{11}I_{12}$	T_1	$I_{11}I_{12}$
11	I_{11}	$I_{14}I_{15}$	T_3	$I_{14}I_{11}$	$I_{11}I_{12}$	T_0	$I_{11}I_{12}$	$I_{12}I_{13}$	T_1	$I_{12}I_{13}$	$I_{13}I_{14}$	T_2	$I_{13}I_{14}$
12	I_{12}	$I_{12}I_{13}$	T_0	$I_{12}I_{13}$	$I_{13}I_{14}$	T_1	$I_{13}I_{14}$	$I_{14}I_{15}$	T_2	$I_{14}I_{15}$	$I_{15}I_0$	T_3	$I_{15}I_{12}$
13	I_{13}	$I_{14}I_{15}$	T_1	$I_{14}I_{15}$	$I_{15}I_0$	T_2	$I_{15}I_0$	I_0I_1	T_3	I_0I_{13}	$I_{13}I_{14}$	T_0	$I_{13}I_{14}$
14	I_{14}	I_0I_1	T_2	I_0I_1	I_1I_2	T_3	I_1I_{14}	$I_{14}I_{15}$	T_0	$I_{14}I_{15}$	$I_{15}I_0$	T_1	$I_{15}I_0$
15	I_{15}	I_2I_3	T_3	I_2I_{15}	$I_{15}I_0$	T_0	$I_{15}I_0$	I_0I_1	T_1	I_0I_1	I_1I_2	T_2	I_1I_2

Figure 4.5 Execution trace for $N = 16$ and $M = 4$

```
procedure MIMD_C1D_1(M);
{MIMD O(1) memory one dimensional convolution}
begin
  pairing (M);
  C1D(p) := 0;
  for j := 0 to M-1 do
  begin
    B(p) := I(p), (igray(p) mod M = M-1-j);
    C1D(p) := C1D(p) + A(p) * T(p);
    MIMDShift (A, -1, N);
    C(p) := B(p); B(p) := A(p); A(p) := C(p); {interchange A and B}
    MIMDShift (T, -1, M);
  end;
end; {of MIMD_C1D_1}
```

Program 4.5 MIMD O(1) memory computation of $C1D$

stages. In the first, we compute the contribution to A and B by all I terms I_j for j even. In the next stage, we do this for the case j odd.

Consider the case $M=8$. If we begin by computing the terms on the major diagonal of Figure 4.6, then PEs (0, 1, 2, ,,, 7) compute $(I_0T_0, I_2T_1, I_4T_2, I_6T_3, I_0T_4, I_2T_5, I_4T_6, I_6T_7)$. The I and T values required by each of the 8 PEs are shown in the first two rows of Figure 4.7. Notice that if we rotate the I values in windows of size 4 by some amount j, then the T values need to be rotated by $2j$ so that each PE has a pair (I, T) whose product is needed in the computation of its A or B value. For this rotation we use the shift sequences F_3 and E_3 defined in Section 2.7. Rotating I by $F[3, 0]$ in size 4 windows and T by $E[3,0]$ in a size 8 window gives the next two rows of Figure 4.7. The result of performing the remaining rotations is also given in Figure 4.7. Figure 4.8 gives the computation of the odd terms.

P_0 $I_0T_0 + I_1T_1 + I_2T_2 + I_3T_3 + I_4T_4 + I_5T_5 + I_6T_6 + I_7T_7$
P_1 $I_1T_0 + I_2T_1 + I_3T_2 + I_4T_3 + I_5T_4 + I_6T_5 + I_7T_6 \,.\, I_0T_7$
P_2 $I_2T_0 + I_3T_1 + I_4T_2 + I_5T_3 + I_6T_4 + I_7T_5 \,.\, I_0T_6 + I_1T_7$
P_3 $I_3T_0 + I_4T_1 + I_5T_2 + I_6T_3 + I_7T_4 \,.\, I_0T_5 + I_1T_6 + I_2T_7$
P_4 $I_4T_0 + I_5T_1 + I_6T_2 + I_7T_3 \,.\, I_0T_4 + I_1T_5 + I_2T_6 + I_3T_7$
P_5 $I_5T_0 + I_6T_1 + I_7T_2 \,.\, I_0T_3 + I_1T_4 + I_2T_5 + I_3T_6 + I_4T_7$
P_6 $I_6T_0 + I_7T_1 \,.\, I_0T_2 + I_1T_3 + I_2T_4 + I_3T_5 + I_4T_6 + I_5T_7$
P_7 $I_7T_0 \,.\, I_0T_1 + I_1T_2 + I_2T_3 + I_3T_4 + I_4T_5 + I_5T_6 + I_6T_7$

Sums to the left of the "." are *A*
Sums to the right of the "." are *B*

Figure 4.6 *A* and *B* values to be computed by each PE

PE	0	1	2	3	4	5	6	7
I	I_0	I_2	I_4	I_6	I_0	I_2	I_4	I_6
T	T_0	T_1	T_2	T_3	T_4	T_5	T_6	T_7
I	I_4	I_6	I_0	I_2	I_4	I_6	I_0	I_2
T	T_4	T_5	T_6	T_7	T_0	T_1	T_2	T_3
I	I_6	I_0	I_2	I_4	I_6	I_0	I_2	I_4
T	T_6	T_7	T_0	T_1	T_2	T_3	T_4	T_5
I	I_2	I_4	I_6	I_0	I_2	I_4	I_6	I_0
T	T_2	T_3	T_4	T_5	T_6	T_7	T_0	T_1

Figure 4.7 Computing the even terms

PE	0	1	2	3	4	5	6	7
I	I_1	I_3	I_5	I_7	I_1	I_3	I_5	I_7
T	T_1	T_2	T_3	T_4	T_5	T_6	T_7	T_0
I	I_5	I_7	I_1	I_3	I_5	I_7	I_1	I_3
T	T_5	T_6	T_7	T_0	T_1	T_2	T_3	T_4
I	I_7	I_1	I_3	I_5	I_7	I_1	I_3	I_5
T	T_7	T_0	T_1	T_2	T_3	T_4	T_5	T_6
I	I_3	I_5	I_7	I_1	I_3	I_5	I_7	I_1
T	T_3	T_4	T_5	T_6	T_7	T_0	T_1	T_2

Figure 4.8 Computing the odd terms

The initial configuration for the I's can be obtained by concentrating the even I's. For this the even I's need to first be ranked. The rank of an even processor is simply half its index. Following the concentration the even I's are only in the left half of the size M window. They can be copied into the right half by a single route along bit $\log_2 M - 1$. Let *RankConcCopy* (I, M) be the procedure that does all this.

The remaining details of the algorithm for one dimensional convolution are provided in Program 4.6, (Ranka and Sahni 1988b). Note that the E's and F's are known only to the control unit. These may be computed, on the fly, in linear time using a stack of height $m = \log M$. The memory required in each hypercube PE is only O(1). Lines 5 through 15 handle the even terms . Notice that $(CShift + 2p)$ **mod** M gives the index of the I value currently in $C(p)$. So, if this index is less than p the term $C * D$ corresponds to the previous block. Otherwise the term $C * D$ is for this PE. The fact that each PE always has a C and a D whose product contributes to either A or B follows from the observations that this is so initially and on each iteration, D rotates twice as much as C. The time complexity of the algorithm is O(M + log$N$).

```
line  procedure C1D_1(M);
1     {O(1) memory SIMD C1D algorithm}
2     begin
3       A(p) := 0; B(p) := 0; m = logM;
4       {even terms}
5       C(p) := I(p); D(p):= T(p);
6       Cshift(p) := 0;
7       RankConcCopy(C, M);
8       for j := 1 to M/2 do
9       begin
10        A(p) := A(p) + C(p) * D(p), ((CShift(p) + 2p) mod mod M ≥ p);
11        B(p) := B(p) + C(p) * D(p), ((CShift(p) + 2p) mod M < p);
12        SIMDShift(C, F[m,j−1], M/2);
13        CShift(p) := (CShift(p) + F[m,j−1]) mod (M/2);
14        SIMDShift(D, E[m,j−1], M);
15      end;
16      {odd terms}
17      C(p) := I(p); D(p) := T(p);
18      SIMDShift(C, −1, M); CShift(p) := 1; SIMDShift(D, −1, M);
19      RankConcCopy(C, M);
20      for j := 1 to M/2 do
21      begin
22        A(p) := A(p) + C(p) * D(p), ((CShift(p) + 2p) mod M ≥ p);
23        B(p) := B(p) + C(p) * D(p), ((CShift(p) + 2p) mod M < p);
24        SIMDShift(C, F[m,j−1], M/2);
```

```
25       CShift (p) := (CShift (p) + F [m,j-1]) mod (M/2);
26       SIMDShift (D, E [m,j-1], M);
27     end;
28     SIMDShift (B, -M, P);
29     C 1D (p) := A (p) + B (p);
30   end; {of C 1D_1}
```

Program 4.6 O(1) memory SIMD algorithm for one dimensional convolution

Chapter 5

Template Matching

5.1 The Problem

The inputs to the image template matching problem are an $N \times N$ image matrix $I[0..N-1, 0..N-1]$ and an $M \times M$ template $T[0..M-1, 0..M-1]$. The output is an $N \times N$ matrix $C2D$ where

$$C2D[i, j] = \sum_{u=0}^{M-1} \sum_{v=0}^{M-1} I[(i+u) \bmod N, (j+v) \bmod N] * T[u,v], \quad 0 \le i, j < N$$

$C2D$ is called the two dimensional convolution of I and T. Template matching, i.e., computing $C2D$, is a fundamental operation in computer vision and image processing. It is often used for edge and object detection; filtering; and image registration (Rosenfeld and Kak 1982, and Ballard and Brown 1985).

5.2 General Square Templates

We shall provide only a high level description of how template matching can be done on SIMD and MIMD hypercubes. At this level of description the steps for both types of hypercubes are the same. In both cases we shall employ the algorithms developed in the previous chapter for one dimensional convolution.

The assumptions and terminology we shall use are summarized below:

(1) $P = N^2$ PEs are available and both N and M are powers of 2.

(2) The N^2 PEs are viewed as an $N \times N$ array as described in Chapter 2. We use (i, j) to refer to the PE in position(i, j) of the $N \times N$ processor array.

(3) $I[i, j]$ is initially in the I register of PE(i, j).

(4) Since N and M are powers of 2, the $N \times N$ array may further be viewed as composed of N^2/M^2 arrays/windows of size $M \times M$. We assume that T is initially in the top left such array/window.

As in the case of one dimensional convolution, we consider two cases for the amount of memory available per processor: O(M) and O(1).

5.2.1 O(M) Memory

When O(M) memory is available, PE(i, j), $0 \le i < N$, $0 \le j < N$ computes M one dimensional convolutions $S(q)$, $0 \le q < M$ defined as

$$S(q) = \sum_{r=0}^{M-1} I((i, (j + r) \bmod N) * T(q, r)$$

Next, $C2D$ is obtained by performing an adjacent sum operation along the columns of the $N \times N$ PE array. Program 5.1 gives the steps that need to be performed to compute $C2D$. The complexity of the resulting algorithm is seen to be O(M^2 + log(N/M)).

---

procedure $C2D\_M(N, M)$;

{Template matching with $O(M)$ memory per PE}

Step 1: Broadcast T to all $M \times M$ windows in the $N \times N$ PE array

Step 2: Perform a data accumulation on I. For this operation, the $N \times N$ PE array is viewed as N independent hypercubes with each row forming one such hypercube. Following the operation, each PE contains the M I values it needs to compute its $S(q)$'s.

Step 3: Compute the $S(q)$'s. Each $S(q)$ is a one dimensional convolution. However, the data accumulation step of the one dimensional convolution algorithms of Chapter 2 may be omitted as the I values have already been accumulated in Step 2. To go from one S to another, the T values need to be circulated along the columns of each $M \times M$ window. This is done using the data circulation algorithms of Chapter 2.

Step 4: Compute $C2D(i, j) = \sum_{r=0}^{M-1} S[r]((i+r) \bmod N, j)$. This is done using the adjacent sum algorithm of Chapter 2 on the columns of the $N \times N$ PE array.

end;

Program 5.1 Two dimensional convolution with each PE having $O(M)$ memory

5.2.2 O(1) Memory

Now, it is not possible for each PE to accumulate the M values of I it needs from its row. Nor is it possible for a PE to compute the values $S(q)$, $0 \le q < M$. We may rewrite the definition of $C2D$ as

$$C2D[i, j] = \sum_{r=0}^{M-1} CXD[i, r, j]$$

where

$$CXD[i, r, j] = \sum_{a=0}^{M-1} I[(i + r) \bmod N, (j + a) \bmod N]*T[r, a]$$

Some of the CXD terms needed for the computation of $C2D(i, j)$ can be computed within the $M \times M$ PE window that contains $PE(i, j)$ as all the needed I and T values are in the window. The remaining terms can be computed by the corresponding PE in the window below it as this window contains the needed I values. Thus each PE computes an E value (for itself) and an F value (for the corresponding PE in the adjacent upper window).

The E and F values are computed in k iterations. During iteration k, the PEs in the k'th row of each window compute their E and F values. These rows have index $k, M+k, 2M+k, \cdots$. Also

$$E(aM + k, j) = \sum_{r=0}^{M-1-k} CXD[aM + k, r, j]$$

and

$$F(aM + k, j) = \sum_{r=M-k}^{M-1} CXD[((a-1)M + k) \bmod N, r, j]$$

For this, we note that PE(i, j) is in the $i \bmod M$ row of the $\lfloor i/M \rfloor$'th window. So, each PE needs to compute

$$A = CXD[\ \lfloor i/M \rfloor M + k,\ i \bmod M - k,\ j\ ] \text{ if } i \bmod M \geq k$$

and

$$B = CXD[\ \lfloor i/M \rfloor M + k - M,\ i \bmod M - k + M,\ j\ ] \text{ if } i \bmod M < k$$

Then, the PEs in rows $aM + k$, $0 \leq a < N/M$ can compute E and F by summing the A's and B's in their column and in their window. Once this has been done, $C2D$ is computed by shifting the F's up the columns by M units and adding to the E's. A high level description of the algorithm is provided in Program 5.2.

The complexity of the algorithm is $O(M^2 + \log N)$ for MIMD hypercubes and $O(M^2 + M\log N)$ for SIMD hypercubes. The SIMD hypercube compexity may be reduced to $O(M + \log N)$ by modifying the $O(1)$ memory C1D algorithm. In this modification, each block of M PEs computes the A and B values for the PEs in its own block (rather than the B values for an adjacent block). This is accomplished by first shifting the I values by one block so that each M block has all the I values it needs. Following this shift, each PE will have two I values. One is the value it had before the shift and the other is the one it received as a result of the shift. Now, using two passes of the C1D algorithm, each PE can compute first its A value and then its B value. As a result the final shift of the B values by one block is eliminated. Since the I values needed in a block of PEs does not change from one iteration of the **for** loop of Step 2 to the next, the initial shifting of I and the *RankConcCopy* steps need to be done just once. Also, the final shifting of the B's is eliminated from the C1D algorithm. As a result, all M invocations of procedure $C1D\_1$ from Step 3 of Program 5.2 take only $O(M^2 + \log N)$ time.

---

procedure $C2D\_1(N, M)$;

{Template matching with O(1) memory per PE}

Step 1: Broadcast T to all $M \times M$ windows in the $N \times N$ PE array

Step 2: Repeat Steps 3 and 4 **for** $k := 0$ **to** $M - 1$

Step 3: PE(i, j) computes $CXD[(\lfloor \frac{i}{M} \rfloor M + k) \bmod N,\ i \bmod M - k,\ j]$ if $i \bmod M \geq k$ using the appropriate O(1) memory one dimensional convolution algorithm and puts the result in A, otherwise $A = 0$;
PE(i, j) computes
$CXD[(\lfloor \frac{i}{M} \rfloor M + k - M) \bmod N,\ i \bmod M - k + M,\ j]$
if $i \bmod M < k$ using the appropriate O(1) memory one dimensional convolution algorithm and puts the result in B, otherwise $B = 0$;

Step 4: Use the data sum operation of Chapter 2, to sum the B's and A's in $PE(\lfloor \frac{i}{M} \rfloor M + k,\ j)$ in F and E respectively. Shift the T values up the columns by 1. The window size for this shift is M.

Step 5: Shift the F values up the columns by M. The window size for this shift is N. $C2D := E + F$.

Program 5.2 Two dimensional convolution with each PE having O(1) memory

5.3 Kirsch Motivated Templates

Kirsch templates, (Ballard and Brown 1985), are commonly used in image processing. Kirsch templates of size 1 ($M = 3$) and 2 ($M = 5$) are shown in Figure 5.1.

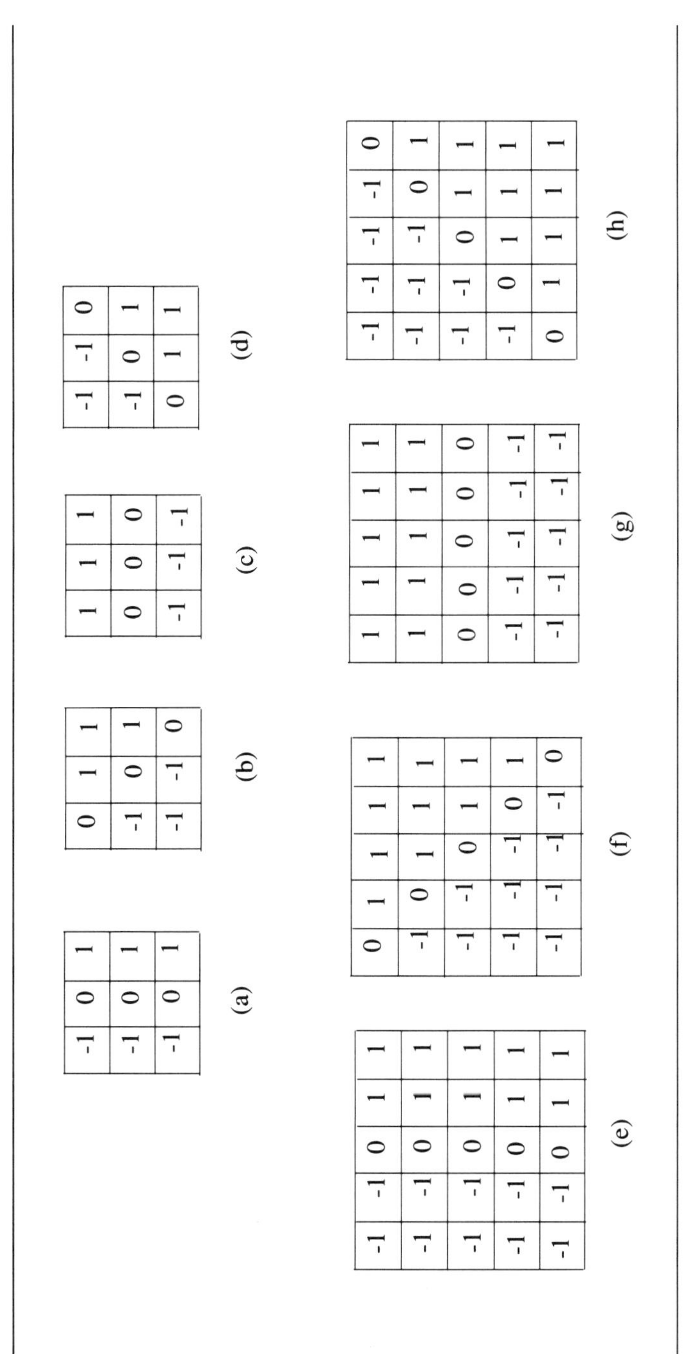

Figure 5.1 Kirsch templates of size 1 ($M = 3$) and 2 ($M = 5$)

By exploiting the special structure of these templates, template matching can be done more efficiently. A high level description of the algorithm is given in Program 5.3. Its complexity is $O(M)$. The amount of memory required per PE is also $O(M)$. While efficient $O(1)$ memory algorithms can also be developed, we shall not do this here as Kirsch templates usually have small M and it is reasonable to assume this much memory is available.

Steps 3, 4, 5, and 6 can be done efficiently by a simple adaptation of the adjacent sum procedure of Chapter 2.

5.4 Medium Grained Template Matching

In the previous sections we have developed algorithms to perform template matching on a fine grain hypercube. Such a computer has the property that the cost of interprocessor communication is comparable to that of a basic arithmetic operation. In this section, we shall consider the template matching problem on a hypercube in which interprocessor communication is relatively expensive and the number of processors is small relative to the image size N. In particular we shall experiment with an NCUBE/7 hypercube which is an MIMD computer capable of having up to 128 processors. However, the computer used for the experiments we report on has only 64 processors. The block diagram for this computer is shown in Figure 5.2. The hypercube is attached to the host computer in a manner akin to the attachment of other peripherals. An NCUBE program consists of a host program together with programs for each of the hypercube processors. The host program loads programs and data onto the hypercube processors. The time to perform a two byte integer addition on each hypercube processor is 4.3 microseconds whereas the time to communicate b bytes to a neighbor processor is approximately $447 + 2.4b$ microseconds.

Several cases of the template matching problem can be studied. These vary in the initial location of the image and the template and the final location of the convolution (result matrix). We consider the following cases . In all of these, the template is initially in the host.

Step 1: Accumulate in A the next M values of I

Step 2: $B[-1] := 0; C[-1] := 0;$

for $i := 0$ **to** $M - 1$ **do**

begin

$B[i] := A[i] + B[i-1];$

$C[i] := A[M-1-i] + C[i-1]$

end;

Do exactly one of the following steps depending on the template type.

Step 3: {Templates of types (a) and (e)}

$$C2D(i, j) = \sum_{a=0}^{M-1} (C[(M-3)/2] - B[(M-3)/2])((i+a) \bmod N, j)$$

Step 4: {Templates of types (b) and (f)}

$$C2D(i, j) = \sum_{a=0}^{M-1} (C[M-2-a] - B[a-1])((i+a) \bmod N, j)$$

Step 5: {Templates of types (c) and (g)}

$$C2D(i, j) = \sum_{a=0}^{(M-3)/2} C[M-1]((i+a) \bmod N, j) - \sum_{(M+1)/2}^{(M-1)} C[M-1]((i+a) \bmod N, j)$$

Step 6: {Templates of types (d) and (h)}

$$C2D(i, j) = \sum_{a=0}^{M-1} (C[a-1] - B[M-2-a])((i+a) \bmod N, j)$$

Program 5.3 Algorithm for Kirsch templates of Figure 5.1

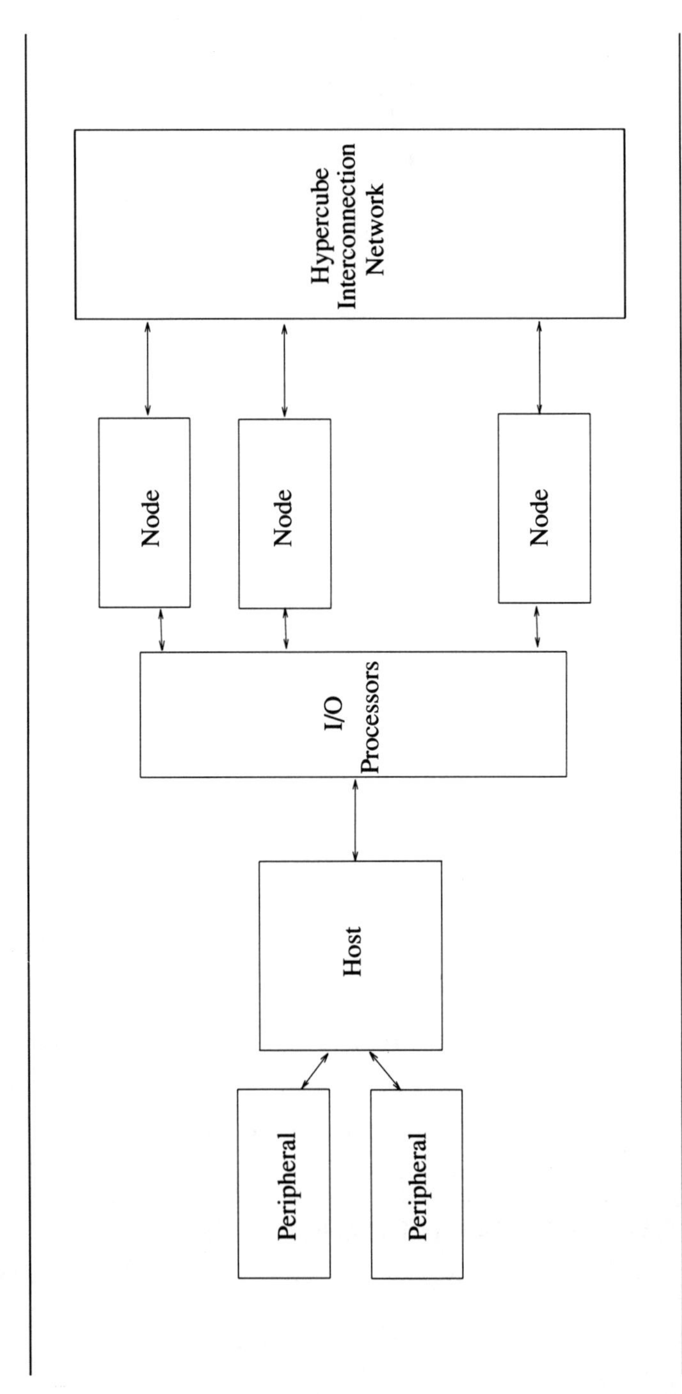

Figure 5.2 NCUBE hypercube computer

1. Host-to-host: The image is in the host initially and the result is to be left in the host also.
2. Hypercube-to-host: The image is initially in the host but the result is left in the hypercube.
3. Hypercube-to-hypercube: The image is initially in the hypercube and the convolution is to be left there too.

Let P be the number of hypercube processors. We assume that P is a perfect square and that $\sqrt{P}$ divides N. Hence, the hypercube may be visualized as a $\sqrt{P} \times \sqrt{P}$ array and the $N \times N$ convolution matrix can be mapped onto this with each processor getting an $N/\sqrt{P} \times N/\sqrt{P}$ block. We assume that each processor has enough memory to hold one copy of the $M \times M$ template. As far as mapping the $N \times N$ image is concerned, we consider the two possibilities:

(1) Overlap Mapping: In this, each processor gets enough of the image to compute all its convolution values. Hence, the processor in position (0, 0) of the mesh gets $I[0 .. n/\sqrt{p} + m - 2, 0 .. n/\sqrt{p} + m - 2]$.

(2) Nonoverlap Mapping: The image is decomposed into $N/\sqrt{P} \times N/\sqrt{P}$ blocks. This is done in the same way as the convolution decomposition. Each processor gets the image block that corresponds to its convolution block.

Notice that if the overlap mapping is used, then the host must transfer more data to each hypercube processor than when the nonoverlap mapping is used. However, no interprocessor communication is needed when the overlap mapping is used. Interprocessor communication is, however, needed when the nonoverlap mapping is used. This can take the form of each processor communicating to its north, east, and northeast neighbor processors the image values they need to compute their convolution. Alternatively, each processor can compute the partial convolution values for its north, northeast, and east neighbors and then communicate these values. In either case, the communication overhead is the same. In our programs, we adopt the latter strategy.

It is also important to note that the communication overhead in the template matching problem is small relative to the computing cost. When the overlap mapping is used, $O(NM\sqrt{P} + PM^2)$ additional data is transmitted from the host to the hypercube nodes (i.e., in addition to the transfer of N^2 image values). However since the host can send data to several nodes in parallel, the overhead penalty is not as severe. While the same amount of data has to be transferred between processors when the nonoverlap mapping is used, the P processors can work in parallel so that the transfer time is approximately that for the transfer of $O(NM/\sqrt{P} + M^2)$ data. In either case, this overhead is expected to be small compared to the time required for the $O(N^2M^2/P)$ computing to be done by each processor.

In each of the three cases listed above, we have assumed that the host broadcasts the template to the hypercube processors using a tree expansion scheme.

The NCUBE/7 run times for $P = 1, 4, 16,$ and 64; $N = 32, 64, 128, 252,$ and 512 and $M = 4, 8, 16,$ and 32 for the overlap memory mapping are given in Figure 5.3 through Figure 5.5. For smaller values of P, the template matching can be done only for small N as there isn't enough memory on a hypercube processor to hold the convolution and the image subblocks assigned to it. The figures show that for the case $N = 512$, $M = 32$, and $P = 64$, the run times for the host-to-host case are approximately 2.6% higher than that for the hypercube-to-host case and approximately 13.0% higher than the hypercube-to-hypercube case. This reflects the cost of transmitting the image and the convolution between the host and the hypercube. The observed speedup is almost equal to the theoretical maximum of P. The speedup and efficiency for $N = 64$ and $M = 8$ are shown in Figure 5.6.

The run times for the nonoverlap mapping are presented only for the hypercube-to-hypercube case. In this case, there are two possibilities:

		M			
P	*N*	4	8	16	32
1	32	0.456	1.479	5.391	20.439
	64	1.832	5.867	21.169	81.485
4	32	0.142	0.383	1.366	5.223
	64	0.524	1.480	5.392	20.440
	128	2.022	5.869	21.170	81.487
16	32	0.104	0.176	0.478	1.596
	64	0.238	0.507	1.477	5.225
	128	0.790	1.754	5.394	20.442
	256	2.925	6.592	21.173	81.491
64	32	0.270	0.421	0.910	2.590
	64	0.428	0.643	1.246	3.172
	128	0.933	1.273	2.349	7.029
	256	2.724	3.293	7.205	22.069
	512	9.365	10.597	25.243	81.491

Times are in seconds
M = template size
N = image size
P = number of processors

Figure 5.3 Overlap mapping: Host-to-host

(1) Overlap of computation and communication between nodes

(2) No overlap of computation and communication between nodes

The experiments indicate that there is no substantial difference in the run times in the above two cases. This is because the amount of computation is much larger than the amount of communication between nodes. The run times for the nonoverlap mapping are given in Figure 5.7. For small template sizes the nonoverlap method is significantly slower

		M			
P	N	4	8	16	32
1	32	0.407	1.308	4.773	18.200
	64	1.600	5.211	18.891	72.268
4	32	0.126	0.355	1.233	4.666
	64	0.462	1.364	4.860	18.226
	128	1.810	5.367	18.974	72.391
16	32	0.069	0.146	0.426	1.483
	64	0.198	0.456	1.402	5.022
	128	0.695	1.643	5.199	18.830
	256	2.620	6.279	19.875	73.350
64	32	0.108	0.190	0.459	1.424
	64	0.200	0.350	0.832	2.533
	128	0.511	0.880	2.111	6.539
	256	1.645	2.786	6.788	21.405
	512	5.968	9.831	24.341	79.440

Times are in seconds

Figure 5.4 Overlap mapping: Hypercube-to-host

than the overlap method. For larger template sizes the difference in run time is not so significant. Much of the difference in the run time is attributable to the following observations:

(1) The program for the nonoverlap case is considerably more complex and so has greater overhead than that for the overlap case.

(2) The data transfer rate from the host to the nodes is much higher than that between nodes.

		M			
P	*N*	4	8	16	32
1	32	0.376	1.274	4.727	18.134
	64	1.504	5.094	18.763	72.105
4	32	0.096	0.320	1.184	4.572
	64	0.378	1.275	4.729	18.136
	128	1.506	5.096	18.764	72.107
16	32	0.028	0.084	0.299	1.146
	64	0.098	0.322	1.185	4.573
	128	0.380	1.277	4.731	18.138
	256	1.508	5.097	18.767	72.109
64	32	0.013	0.027	0.086	0.291
	64	0.030	0.086	0.301	1.148
	128	0.100	0.324	1.187	4.575
	256	0.381	1.279	4.733	18.139
	512	1.510	5.099	18.768	72.110

Times are in seconds

Figure 5.5 Overlap mapping: Hypercube-to-hypercube

(3) For larger template size the computation time significantly dominates the communication time.

Figure 5.8 shows the time required by a single processor CRAY-2 supercomputer to perform template matching. These are approximately one fifth of the hypercube-to-hypercube times on the NCUBE/7 with 64 processors.

	P	1	4	16	64
Host-to-host	Speed up	1.00	3.96	11.57	9.12
	Efficiency	1.00	0.99	0.72	0.14
Hypercube-to-host	Speed up	1.00	3.82	11.43	14.89
	Efficiency	1.00	0.95	0.71	0.23
Hypercube-to-hypercube	Speed up	1.00	3.99	15.82	59.23
	Efficiency	1.00	0.998	0.99	0.93

Times are in seconds

Figure 5.6 Overlap mapping: speedup and efficiency for $N = 64$ and $M = 8$

		M			
P	N	4	8	16	32
1	32	0.505	1.857	7.000	20.450
4	32	0.139	0.482	1.417	
	64	0.514	1.872	7.026	20.497
16	32	0.045	0.115		
	64	0.142	0.484	1.422	
	128	0.516	1.874	7.031	20.510
64	32	0.021			
	64	0.047	0.118		
	128	0.144	0.487	1.426	
	256	0.519	1.878	7.036	20.520

Times are in seconds

Figure 5.7 Nonoverlap mapping: Hypercube-to-hypercube

	M			
N	4	8	16	32
64	0.007	0.023	0.086	0.345
128	0.022	0.080	0.300	1.205
256	0.073	0.283	1.118	4.485
512	0.273	1.082	4.273	17.350

Times are in seconds

Figure 5.8 Template matching on CRAY-2

Chapter 6

Hough Transform

6.1 Introduction

Hough transforms are used to detect straight lines or edges in an image. Let L be a straight line in the x–y plane. The *normal* to this line is another straight line that is perpendicular to it and passes through the origin (0, 0) (Figure 6.1). Let θ be the angle the normal makes with the x-axis and let r be the length of the normal. All points (x_i, y_i) on L satisfy the equality

$$x_i \cos\theta + y_i \sin\theta = r$$

The Hough transform utilizes this equality to detect straight lines or edges in an image. We try out a set $\{\theta_j \mid 0 \le j < p\}$ of p possible values for the angle θ. This is equivalent to trying out a set of p possible slopes for the lines being detected. A θ_j and a point (x_i, y_i) together uniquely define a line L through this point. The length of the normal to this line is given by the above equality. Furthermore, for any θ_j, all image points (x_i, y_i) that have the same normal length, $x_i\cos\theta_j + y_i\sin\theta_j$, lie on the same line L. (This line is uniquely defined by the angle, θ_j, and the length of the normal.) For each θ_j, we determine the number of image points that have the same normal length. By knowing how many image points have

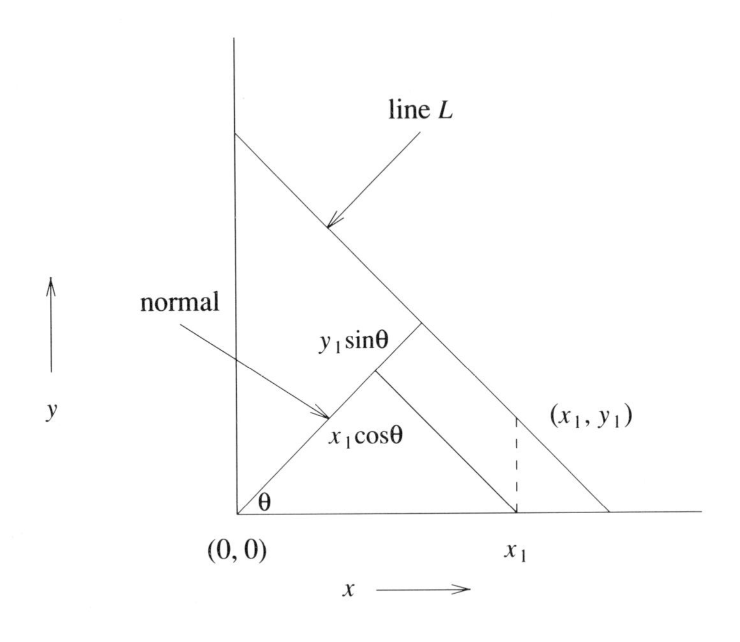

Figure 6.1 A line L and its normal

the same normal (i.e., same normal length and same normal angle) we can determine the probabilty that the image has a line with that normal.

Let $I[0..N-1, 0..N-1]$ be an $N \times N$ image such that $I[x, y] = 1$ iff the image point $[x, y]$ is a possible edge point. $I[x, y] = 0$ otherwise. The p angle Hough transform of I is the array H such that

$$H[i, j] = |\{(x, y) | i = \lfloor x\cos\theta_j + y\sin\theta_j \rfloor, \theta_j = \frac{\pi}{p}(j+1) \text{ and } I[x, y] = 1\}|.$$

j takes on the integer values $0, 1, \cdots, p-1$. These correspond to the p angles $\theta_j = \frac{\pi}{p}(j+1), 0 \le j < p$. Hence $0 < \theta_j \le \pi$. For θ_j in this range and x and y in the range $[0, N-1]$, $\lfloor x\cos\theta_j + y\sin\theta_j \rfloor$ is in the range $[-\sqrt{2}N, \sqrt{2}N]$. Hence H is at most a $2\sqrt{2}N \times p$ matrix.

We shall explicitly consider the computation of $H(i, j)$ only for $i > 0$. The computation for the case $i \leq 0$ is similar. Hence i is in the range $[0, \sqrt{2}N)$ and j is in the range $[0, p)$. We assume that N is a power of two and that $2N^2$ PEs are available. These are viewed as an $N \times 2N$ array as discussed in Chapter 1 for SIMD and MIMD hypercubes. Actually, only $N \times \sqrt{2}N$ PEs are needed; however, a hypercube must have a power of 2 processors. Furthermore, it is assumed that p divides N.

The image pixel $I[i, j]$ is initially stored in PE (i, j) $0 \leq i, j < N$ in the above array view. $H[i, j]$ is stored in PE (i, j) on completion. The hypercube algorithms we develop are conceptually similar to those developed by Cypher and Sanz (1987) for mesh connected computers. The development here is from (Ranka and Sahni 1989a).

6.2 MIMD Algorithm

For simplicity, we divide the computation of $H[i, j]$, $i > 0$, $0 \leq j < p$ into four parts. These, respectively, correspond to the cases $0 < j < p/4$, $p/4 \leq j < p/2$, $p/2 \leq j < 3p/4$, and $3p/4 \leq j < p$. First, consider the case $p/4 \leq j < p/2$. Now, $\pi/4 < \theta_j \leq \pi/2$. The following two lemmas suggest a computational scheme for this case.

Lemma 6.1 When $\pi/4 < \theta_j \leq \pi/2$, two pixels (x, y) and $(x, y + z)$, $z > 0$, can contribute to the count of the same $H[i, j]$ only if $z = 1$.

Proof: If (x, y) and $(x, y + z)$ both contribute to the count of $H[i, j]$, then

$$i = \lfloor x\cos\theta_j + y\sin\theta_j \rfloor = \lfloor x\cos\theta_j + (y + z)\sin\theta_j \rfloor$$

for some j, $p/4 \leq j < p/2$. Hence

$$(y + z)\sin\theta_j - y\sin\theta_j < 1$$

or

$$z\sin\theta_j < 1$$

Since $\pi/4 < \theta_j \leq \pi/2$, $\sin\theta_j > \sin \pi/4 > 0{\cdot}5{\cdot}$ Since z is a positive integer, only $z = 1$ can satisfy the relation $z\sin\theta_j \leq 1$. □

Lemma 6.2 When $\pi/4 < \theta_j \leq \pi/2$, two pixels (x, y) and $(x + 1, z)$ can contribute to the count of the same $H[i, j]$ only if $z \in \{y, y-1\}$.

Proof: If (x, y) and $(x + 1, z)$ contribute to the same $H[i, j]$, then $i = \lfloor x\cos\theta_j + y\sin\theta_j \rfloor = \lfloor (x + 1)\cos\theta_j + z\sin\theta_j \rfloor$.
So,

$$|(x + 1)\cos\theta_j - x\cos\theta_j + (z-y)\sin\theta_j| \leq 1$$

or

$$|\cos\theta_j + (z-y)\sin\theta_j| \leq 1$$

or

$$|cot\theta_j + (z-y)| \leq cosec\,\theta_j$$

or

$$-cosec\,\theta_j - cot\theta_j \leq z-y \leq cosec\,\theta_j - cot\theta_j$$

Since y and z are integers and θ_j is in the above range, it follows that $-1 \leq z-y \leq 0$. Hence $z \in \{y, y-1\}$. □

The computation of $H[i, j]$ for $i > 0$ and $\pi/4 \leq \theta_j < \pi/2$ can be done in two phases. In the first, subhypercubes of size $p{\times}2N$ compute

$$h[i, j] = |\{(x, y) \mid i = \lfloor x\cos\theta_j + y\sin\theta_j \rfloor, \pi/4 \leq \theta_j < \pi/2, I[x, y] = 1, \text{ and } (x, y) \text{ is in this subhypercube}\}.$$

In the second phase, the $h[i, j]$ values from the different subhypercubes are summed to get

$$H[i, j] = \sum_{subhypercubes} h[i, j], i > 0, p/4 \leq j < p/2.$$

The phase 1 algorithm for each PE in a $p \times 2N$ subhypercube is given in Program 6.1. In this algorithm, $[x, y]$ denotes a PE index relative to the whole $N \times 2N$ hypercube and $[w, y]$ denotes the index of the same PE relative to the $p \times 2N$ subhypercube it is in. Note that $w = x \bmod p$.

The h values are computed in a pipeline manner. The PEs in row 0 of a $p \times 2N$ subhypercube initiate a record $Z = (i, j, \text{sine}, \text{cosine}, q)$ such that $h[i, j] = q$ is the number of pixels on this row that contribute to $h[i, j]$. This is done by first computing i for each pixel in row zero (line 7) for a fixed $j = p/2 - l - 2$. Lemma 6.1 is used in lines 23-25 to combine records that represent the same $h[i, j]$ entry. This row of Z records created in row zero moves down the $p \times 2N$ subhypercube one row per iteration (line 26). Lines 11-21 update the row of Z values received. Each such row corresponds to a fixed j. For this j, PE$[w, y]$ determines the h entry $[i', j]$ it is to contribute to (line 14). If this is the same entry as received from PE$[w-1, y]$ then the two are added together. If $i = \phi$ for the received entry, then $[i', j]$ can occupy this Z space. If $i \neq \phi$, then from Lemma 3.2 we know that Z can combine only with the new entry $[i', j]$ of PE$[w, y-1]$.

Following the iteration $l = 5p/4 - 1$, the last initiated row (i.e., $j = p/4$) has passed through row $p-1$ of the $p \times 2N$ subhypercube. At this time, the PEs in row r of the subhypercube contain records with $j = p/4 + r$, $0 \leq r < p/4$. The records in each row may be reordered such that the record in PE$[w, y]$ has $y = i$ by performing a random access write. Because of the initial ordering of i values in a row, this random access write can be performed in $O(\log N)$ time by eliminating the sort step.

The phase 2 summing of the $h[i, j]$ values is now easily done in $O(\log N)$ time using window sum. Since the phase 1 algorithm of Program 6.1 only shifts by 1 along columns and/or rows, each iteration of this

```
1 for l := 0 to 5p/4–1 do
2 begin
3   if (w = 0) and (l < p/4) then
4   begin {row 0 initiates next θ_j}
5      create a record Z = (i, j, sine, cosine, q) with
6      sine = sin(θ), cosine = cos(θ), where θ = (π/p)(p/2 – l + m);
7      i = ⌊x cosine + y sine⌋; j = p/2–l–1;  q = I[x, y];
8   end
9   else begin
10        if max{1, l–p/4 + 1} ≤ w ≤ l and y < N
11        then begin {add in this PE's contribution}
12              Let Z be the record received from PE (w–1, y);
13              Let i′ = ⌊x cosine + y sine⌋ and q′ = I[x, y];
14              if i = i′ then set q = q + q′
15              else if i = φ then set i = i′ and q = q′
16                   else send q to PE(x, (y–1) mod 2N)
                          and set Z = (i′, j, sine, cosine, q′)
17              if a q is received from PE (x, y + 1)
                  update own q to q + received q
18            end
19        else if y > N and a Z is received from PE (x, (y + 1) mod 2N)
20             then send old Z (if any) to PE on left
21      end;
22     {combine records with same (i, j) values}
23     if (⌊x cosine + y sine⌋ = ⌊x cosine + (y–1) sine⌋) and (0 < y < N)
24     then send h to PE[x, y–1] and set i = φ
25     else if a q value is received set q = q + received q;
26     send Z to PE ((w + 1) mod p, y);
27 end;
```

Program 6.1 MIMD algorithm

algorithm takes only O(1) time. Hence the complexity of the phase 1 algorithm is $O(p)$. The overall time needed to compute H for $p/4 \le x < p/2$ is therefore $O(p + \log N)$.

The remaining three cases for j are done in a similar way. Actually, the four cases need not be computed independently as suggested above. In particular, all the computation following phase 1 can be done in parallel for all the cases.

6.3 SIMD Algorithms

We develop two $O(p + \log N)$ SIMD hypercube algorithms. One uses $O(\log N)$ memory per PE while the other uses O(1). The O(1) memory algorithm is slightly more complex than the $O(\log N)$ memory one. Both algorithms are adaptations of our MIMD algorithm. The computations following phase 1 (Program 6.1) are easily performed in $O(\log N)$ time on an SIMD hypercube using O(1) memory per PE. So we concentrate on adapting phase 1. The phase 1 algorithm performs $O(p)$ unit shifts along rows and columns of $p \times 2N$ subhypercubes. In an SIMD hypercube, each such row shift takes $O(\log N)$ time while each unit column shift takes $O(\log p)$ time. So a direct simulation of phase 1 takes $O(p \log(Np))$ time.

6.3.1 $O(\log N)$ Memory Per PE

In this case, we divide the $5p/4$ iterations of the **for** loop of Program 6.1 into blocks of $\log N$ consecutive iterations. In each such block, a Z record initially in PE$[x, y]$ can be augmented by pixel values in PEs $[x + l, y-m]$, $0 \le l < \log N$, $-1 \le m < \log N$. To avoid unit shifts along the rows, each PE$[q,r]$ begins by accumulating the pixel value in PE$[q, r-m]$, $-1 \le m < \log N$. Now it is necessary to route the Z records only down a column; i.e., a Z record initially in PE $[x, y]$ needs to visit PEs $[x + l, y]$, $0 \le l < \log N$. These PEs contain the pixel values needed to update Z to its values following the block of iterations in Program 6.1. This routing is done using the circulation algorithms in windows of size $\log N$ rather than by unit shifts. The initial pixel accumulation takes $O(\log N)$ time and the circulation and Z updates also take $O(\log N)$ time.

Following the circulation, the Z records return to their originating PEs and need to be routed left and down by a distance of $O(\log N)$. This can be accomplished in $O(\log N)$ time on an SIMD hypercube. In this way, we are able to simulate $O(\log N)$ iterations of the MIMD algorithm in $O(\log N)$ time on an SIMD hypercube. Hence the overall asymptotic run time of the SIMD simulation is the same as that of the original MIMD algorithm.

6.3.2 O(1) Memory Per PE

When $\log^2 N/p \leq c$ for some constant, a careful analysis shows that using the strategy employed in the $O(\log N)$ memory algorithm, the memory requirements can be reduced to $O(1)$. In any $\log N$ block of iterations, two pixels $[x, y]$ and $[w, z]$ contribute to the same Z record only if

$$\lfloor x\cos\theta + y\sin\theta \rfloor = \lfloor w\cos\theta + z\sin\theta \rfloor$$

Since $w \leq x + \log N - 1$ during the $\log N$ iterations, we get

$$|(\log N - 1)\cos\theta + (z - y)\sin\theta| \leq 1$$

or

$$-cosec\,\theta \leq (\log N - 1)cot\,\theta + z - y \leq cosec\,\theta$$

or

$$-cosec\,\theta - (\log N - 1)cot\,\theta \leq z - y \leq cosec\,\theta - (\log N - 1)cot\,\theta$$

For any fixed $\theta \in \{\pi/4, \pi/2\}$,

$$z \in [y - (\log N - 1)cot\,\theta - cosec\,\theta,\ y - (\log N - 1)cot\,\theta + cosec\,\theta]$$

or

$$z \in [y - (\log N - 1)cot\,\theta - \sqrt{2},\ y - (\log N - 1)cot\,\theta + \sqrt{2}]$$

There are only a constant number of integers in this range. During a $\log N$ block of iterations, Z records with j value differing by up to $\log N - 1$ may pass through a given PE. This corresponds to a θ variation from θ_1 to θ_2 where $\theta_2 - \theta_1 = \frac{\pi}{p}(\log N - 1)$.

Hence the leftmost column from which a contributing pixel is required has a maximum range of

$$cosec\,\theta_1 + (\log N - 1)cot\,\theta_1 - cosec\,\theta_2 - (\log N - 1)cot\,\theta_2$$

$$\leq cosec\,\theta_1 - cosec\,\theta_2 + (\log N - 1)(cot\,\theta_1 - cot\,\theta_2)$$

$$\leq cosec\,\pi/4 + (\log N - 1)\frac{\cos\theta_1 \sin\theta_2 - \cos\theta_2 \sin\theta_1}{\sin\theta_1 \sin\theta_2}$$

$$< cosec\,\pi/4 + 2(\log N - 1)sin\,(\theta_2 - \theta_1)$$

$$< cosec\,\pi/4 + 2(\log N - 1)(\theta_2 - \theta_1)$$

$$= cosec\,\pi/4 + 2(\log N - 1)(\log N - 1)\pi/p$$

$$< cosec\,\pi/4 + 2\pi c$$

Hence each PE need accumulate only a constant number of pixels from its row rather than the $O(\log N)$ pixels being accumulated in the $O(\log N)$ memory algorithm. This accumulation is done in $O(\log N)$ time. The run time is the same as that of the $O(\log N)$ memory algorithm, but the memory requirements are reduced to $O(1)$.

6.4 NCUBE Algorithms

6.4.1 Two NCUBE Algorithms

We view the P hypercube nodes as forming rings (Chapter 1). For any node i, let left (i) and right (i), respectively, be the node counter-clockwise and clockwise from node i. Let logical (i) be the logical index of node i in the ring. The $N \times N$ image array is initially distributed over the nodes with each node getting an $N \times N/p$ block. Logical node 0 gets the first block, logical node 1 the next block, and so on. Similarly, on completion, the $2\sqrt{2}N \times p$ Hough array H is distributed over the nodes in blocks of size $2\sqrt{2}N \times p/P$. We assume that the number of hypercube nodes P divides the number of angles p as well as the image dimension N. It is further assumed that the thresholding function has already been applied to the pixels and each node has a list of pairs (x, y) such that $I[x, y]$ passes the threshold. We call this list the edge list for the node.

Our first algorithm is given in Program 6.2. This algorithm runs on each hypercube node. As remarked earlier, each node has an edge list and an H partition. The H partitions move along the ring one node at a time. When an H partition reaches any node, the edge list of that node is used to update it, accounting for all contributions these edges make to this H partition. Procedure Update*H*Partition does precisely this. *jBegin* is the j value corresponding to the first angle (column) in the H partition currently in the node. $size = p/P$ is the number of columns in an H partition.

In the algorithm of Program 6.2 no attempt is made to overlap computation with communication. Following the send of an H partition to its right neighbor, the node is idle until the receive of the H partition from its left neighbor is complete. Figure 6.2 shows the activity of a node as a function of time.

During the compute phase, an H partition is updated. Let t_c be the time needed to do this. Let t_t be the time for an H partition to travel from a sending node to its destination node. So t_t is the elapsed time between

```
procedure UpdateHpartition (H);
begin
   for each (x, y) in edge list  do
      for (j: = jBegin  to jBegin + size−1  do
      begin
         θ = (π/p)(j + 1);
         i = xcosθ + ysinθ;
         increment H [i,theta ] by 1;
      end;
end;  {of UpdateHpartition}

l: =  logical index of this node, size:= p/P;
jBegin: = size*l;
initialize own H partition to zero;
for i: = 0 to P−1 do
begin
   UpdateHpartition ;
   send own H partition to node on right;
   receive H partition from node on left;
   jBegin := (jBegin  − size) mod p;
end;
```

Program 6.2 Non-overlapping algorithm to compute H

the initiation of the transfer and the receipt of the partition. The time required by the nonoverlapping algorithm of Program 6.2 is $P(t_c + t_t)$.

Our second algorithm, (Program 6.3), attempts to overlap as much of the transmission time t_t with computation. This, unfortunately, results in an increase in the computation time as some additional work is to be done. At the end of each iteration of the *for* loop, the H partition in a node l is sent to the node on its right. The next iteration proceeds while the H partition is in transit. For this, a temporary space T of the same size as H is used to accumulate the contribution of this node's edge list to the

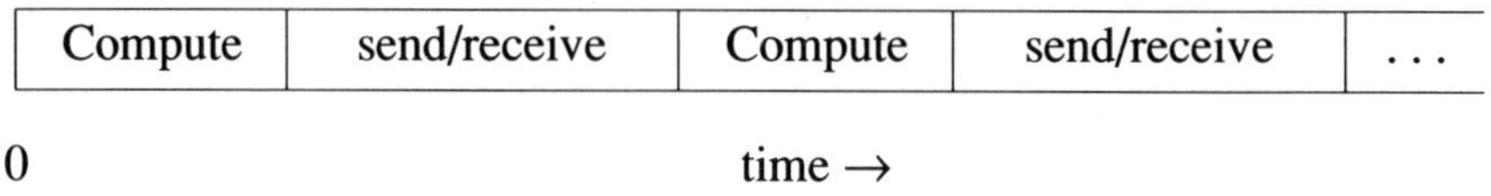

Compute	send/receive	Compute	send/receive	...

0 time →

Figure 6.2 Node activity using Program 6.2

```
l := logical index of this node; size = p/P;
jBegin := size*l;
for i := 0 to P-1 do
begin
   if i = 0 then
   begin
      initialize own H partition to zero;
      UpdateHPartition(H);
   end
   else
   begin
      initialize T to zero;
      Update H Partition (T);
      Receive H Partition from left (l);
      H := H + T;
   end;
   send H to right (l);
   jBegin := (jBegin - size) mod p
end;
end;
```

Program 6.3 Overlapping algorithm for H

H partition it has yet to receive from its left neighbor. Following this computation, the received H portion and T are added as the resulting H partition transmitted to the right.

Relative to the nonoverlapping algorithm, the overlapping algorithm does $P-1$ initializations of T and executions of $H := H + T$ extra computational work. Let t_{init} be the time to initialize T and t_{add} the time to execute $H := H + T$. If $t_t \leq t_{init} + t_c$, the time diagram has the form shown in Figure 6.3 (a). The overall time for the algorithm is $Pt_c + (P-1)(t_{init} + t_{add}) + t_t$ when $t_t \leq t_{init} + t_c$. So if $t_{init} + t_{add} < t_t$, the overlapping algorithm will outperform the nonoverlapping algorithm.

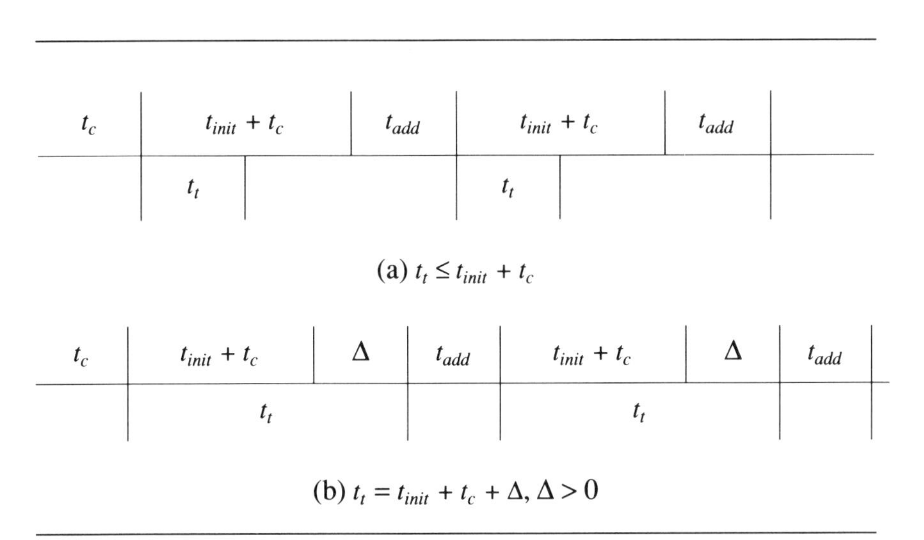

Figure 6.3 Node activity as a function of time

When $t_t = t_{init} + t_c + \delta, \delta > 0$, the time diagram is as in Figure 6.3 (b). In this case, the algorithm run time is $t_c + (P-1)t_{add} + Pt_t = Pt_c + (P-1)(t_{init} + t_{add} + \delta) + t_t$. For the overlapping algorithm to outperform the nonoverlapping algorithm, we need $t_{init} + t_{add} + \delta < t_t$.

6.4.2 Load Balancing

The preceding analysis is somewhat idealistic as it assumes that t_c is the same in each node. Actually, the size of the edge list in each node is different and this difference significantly impacts the performance of the algorithm. The node with the maximum number of edges becomes a bottleneck. To reduce the run time, one may attempt to obtain an equal or near equal distribution of the edges over the P nodes. Note that even though the image matrix I is equally distributed over the nodes, the edge lists may not be, as a different number of pixels in each I partition will pass the threshold. We shall use the term *load* to refer to the number of pixels in a node that passes this threshold. I.e., load is the size of the nodes' edge list. Two heuristics to balance the load are given in Program 6.4 and Program 6.5. In both, load balancing is accomplished by averaging over the load in processors that are directly connected. The variables used have the following significance

MyLoad = current load in the node processor
HisLoad = load in a directly connected node processor
MyLoadSize = size of the load in the node processor
HisLoadSize = size of the load in a directly connected node processor
avg = average size of the load of the two processors

The only difference between the two variations is that in the first one a processor transmits its entire work load (including the necessary data) to its neighbor processor, while in the second variation only the amount in excess of the average is transmitted. However, in order to achieve this reduction in load transmission, it is necessary to first determine how much of the load is to be transmitted. This requires an initial exchange of the load size. Hence variation 2 requires twice as many message transmissions. Each message of variation 2 is potentially shorter than each message transmitted by variation 1. We expect variation 1 to be faster than variation 2 when the number of bytes in *MyLoad* and *HisLoad* is relatively small and the time to set up a data transmission relatively large. Otherwise, variation 2 is expected to require less time.

```
procedure LoadBalance1;
begin
  for i: = 0 to CubeSize do
  begin
    Send MyLoad to neighbor processor along dimension i;
    Receive HisLoad from neighbor processor along dimension i
    and append to Myload;
    avg = (MyLoadSize + HisLoadSize + 1)/2;
    if (MyLoadSize > Avg) then MyLoadSize = Avg
    else if (HisloadSize > Avg)
        then MyLoadSize := MyLoadSize + HisLoadSize - Avg;
  end;
end; {of LoadBalance 1}
```

Program 6.4 First load balancing heuristic

6.4.3 Experimental Results

The nonoverlapping and overlapping algorithms of Section 6.4.1, as well as the load balancing heuristics of Section 6.4.2 were programmed in C and run on an NCUBE/7 hypercube with 64 nodes. Experiments were conducted using randomly generated images of size $N \times N$ for $N = 32$, 64, 128, 256, and 512. The percentage of pixels in an $N \times N$ image that passed the threshold was fixed at 5%, 10%, or 20%. The number of edge pixels in each nodes I partition was determined using a truncated normal distribution with variance being one of 4%, 10%, and 64% of the mean. In all cases, p was equal to 180.

The run time of the two load-balancing heuristics was approximately the same, with the second heuristic having a slight edge. Furthermore, the time to load balance is less than 2% of the overall run time (load balance followed by Hough transform computation). The run time of the nonoverlapping algorithm, both with and without load balancing, is given in Figure 6.4 through Figure 6.6 for the cases of $P = 4$, 16, and 64, respectively. We see that as the load variance increases from 4% to

```
procedure LoadBalance 2;
begin
  for i : = 0 to CubeSize do
  begin
    Send MyLoadSize to neighbor processor along dimension i;
    Receive HisLoadSize from neighbor processor along dimension i;
    avg := (MyLoadSize + HisLoadSize + 1)/2;
    if (MyLoadSize > Avg) then
    begin
      Send extra load (MyLoadSize -Avg) to neighbor processor
      along dimension i;
      MyLoadSize := Avg;
    end
    else if(HisLoadSize > Avg) then
    begin
      Receive extra load (Avg -HisLoadSize) from neighbor
      processor along dimension i;
      MyLoadSize := MyLoadSize + HisLoadSize - Avg;
    end;
  end;
end; {of LoadBalance 2}
```

Program 6.5 Second load balancing heuristic

64%, the run time of the nonoverlapping algorithm without load balancing increases significantly. In fact, it almost doubles. With load balancing, however, the run time is quite stable. Furthermore, it is always less than the run time for 4% variance without load balancing. When the variance in load is 64%, load balancing results in a 25% to 53% reduction in run time!

Note that the average load per node when $P = 4$ and $N = 128$ is the same as when $P = 16$ and $N = 256$ and when $P = 64$ and $N = 512$. From the experimental data we see that the run time remains virtually unchanged as P increases, provided the load per node is unchanged. Hence the algorithm scales well.

The run times for the overlapping algorithm with load balancing are given in Figure 6.7. These times are generally slightly larger than those for the nonoverlapping algorithm with load balancing. So, the computational overhead introduced by the overlapping algorithm more or less balances the positive effects of overlapping computation and communication. For comparison purposes, the run times on a single hypercube node are given in Figure 6.8 for the cases $N = 16$, 32 and 64. The case $N = 128$ could not be run for lack of sufficient memory.

Figure 6.9 gives the speedup and efficiency figure achieved by the nonoverlapping algorithm with load balancing for the cases: variance = 64%, *%edges* = 20, and $N = 64$ and 128.

Image Size	%	No Load Balancing			Load Balance 2		
		Variance					
	edges	4 %	16 %	64 %	4 %	16 %	64 %
32×32	5	0.2802	0.3138	0.3940	0.2819	0.2785	0.2804
	10	0.5627	0.6035	1.1563	0.5531	0.5527	0.5527
	20	1.1439	1.3364	1.7874	1.0976	1.0956	1.0967
64×64	5	1.1465	1.3044	1.7575	1.1202	1.1187	1.1176
	10	2.2428	2.4485	3.4152	2.1878	2.1853	2.1818
	20	4.4974	4.7970	8.0548	4.3171	4.3238	4.3190
128×128	5	4.4966	5.0359	7.8626	4.3605	4.3550	4.3564
	10	8.9968	10.0017	15.5813	8.6474	8.6393	8.6423
	20	18.0087	19.1119	31.7456	17.2247	17.2349	17.2108

Figure 6.4 Number of nodes = 4, no overlap

Image Size	%	No Load Balancing			Load Balance 2		
		Variance					
	edges	4 %	16 %	64 %	4 %	16 %	64 %
64 × 64	5	0.2964	0.3494	0.5622	0.2981	0.3012	0.2922
	10	0.5949	0.6803	1.1556	0.5927	0.5830	0.5712
	20	1.1827	1.4140	2.0260	1.1615	1.1574	1.1313
128 × 128	5	1.2088	1.4113	2.2415	1.1915	1.1798	1.1570
	10	2.3558	2.7429	5.3075	2.3256	2.3065	2.2469
	20	4.6616	5.3293	9.0813	4.5909	4.5600	4.4518
256 × 250	5	4.6854	5.6429	9.3724	4.6283	4.5810	4.4624
	10	9.3130	11.0024	18.0237	9.1721	9.1270	8.8296
	20	18.4712	21.4809	33.9781	18.2738	18.1359	17.6917

Figure 6.5 Number of nodes = 16, no overlap

Image Size	%	No Load Balancing			Load Balance 2		
		Variance					
	edges	4 %	16 %	64 %	4 %	16 %	64 %
128 × 128	5	0.3462	0.4200	0.6449	0.3416	0.3481	0.3400
	10	0.6512	0.7735	1.3975	0.6313	0.6315	0.6232
	20	1.2692	1.5239	2.8156	1.2051	1.2062	1.1960
256 × 256	5	1.2638	1.5371	2.7062	1.2291	1.2229	1.2057
	10	2.4770	2.9324	5.1395	2.3543	2.3476	2.3288
	20	4.9057	6.0051	11.4614	4.6170	4.6020	4.5470
512 × 512	5	4.9077	5.8094	10.8207	4.6485	4.6232	4.5784
	10	9.7492	11.7256	20.7631	9.1908	9.1611	9.0623
	20	19.3672	23.9020	38.0617	18.2782	18.2166	18.0306

Figure 6.6 Number of nodes = 64, no overlap

Block Size	% edges	$P = 4$ 4 %	16 %	64 %
32×32	5	0.3704	0.3689	0.3708
	10	0.6424	0.6425	0.6925
	20	1.1862	1.1851	1.1857
64×64	5	1.3030	1.3011	1.3009
	10	2.3686	2.3680	2.3614
	20	4.4927	4.4967	4.4956
128×128	5	4.7045	4.6999	4.7019
	10	8.9787	8.9760	8.9835
	20	17.5374	17.5426	17.5304

Figure 6.7

Image Size	% edges	Time in Seconds
16×16	5	0.3005
	10	0.5636
	20	1.1016
32×32	5	1.1597
	10	2.2209
	20	4.3527
64×64	5	4.4399
	10	8.7194
	20	17.2660

Figure 6.8 Number of nodes = 1

edges	no. of Nodes	Image = 64 × 64 Time	Speedup	Image = 128 × 128 Time	Est.Speedup
5	1	3.8603	1.0000		
	4	0.9787	3.9440	0.3964	3.9440
	16	0.2844	13.5728	1.0734	14.5640
	64	0.1551	24.8754	6.3414	45.7945
10	1	7.6151	1.0000		
	4	1.91169	3.9724	8.3301	3.9724
	16	0.5263	14.4682	2.2187	14.9288
	64	0.2046	37.2515	0.6278	52.7058
20	1	15.6470	1.0000		
	4	3.9246	3.9868	17.0167	3.9868
	16	1.0529	14.8604	4.4732	15.1660
	64	0.3374	46.3741	1.1777	56.6400

Figure 6.9 No overlap between communication/computation Variance of edges = 64%

Chapter 7

Clustering

7.1 Introduction

A *feature vector* v is a set of measurements $(v_1, v_2, \ldots, v_M)$ which map the important properties of an image into a Euclidean space of dimension M (Ballard and Brown 1985). *Clustering* partitions a set of feature vectors into groups. It is a valuable tool in exploratory pattern analysis and helps making hypotheses about the structure of data. It is important in syntactic pattern recognition, image segmentation and registration. There are many methods for clustering feature vectors (Ballard and Brown 1985, Duda and Hart 1973, Fukunaga 1972, Fu 1974, Rosenfeld and Kak 1982, and Tou and Gonzalez 1974). One popular technique is *squared error* clustering.

Let N represent the number of patterns which are to be partitioned. Each pattern has M features (M is usually less than N). Let $F[0..N-1, 0..M-1]$ be the feature matrix such that F[i, j] denotes the value of the j'th feature in the i'th pattern. Let $S_1, S_2, \ldots, S_K$ be K clusters. Each pattern belongs to exactly one of the clusters. Let $C[i]$ represent the cluster to which pattern i belongs. Thus, we can define S_k as

$$S_k = \{i \mid C[i] = k\}, 0 \le k < K$$

Further, $| S_k |$ is the cardinality or size of the partition S_k. The *center* of cluster k is a $1 \times M$ vector defined as:

$$center[k, j] = \frac{1}{|S_k|} \sum_{i \in S_k} F[i, j], \quad 0 \leq j < M$$

The *squared distance* $d2$ between pattern i and cluster k is

$$d2[i, k] = \sum_{j=0}^{M-1} (F[i, j] - center[k, j])^2$$

The *squared error* for the k'th cluster is defined as

$$E2[k] = \sum_{i \in S_k} d2[i, k], \quad 0 \leq k < K$$

and the *squared error for the clustering* is

$$ERROR[K] = \sum_{k=0}^{K-1} E2[k]$$

In the clustering problem, we are required to partition the N patterns such that the squared error for the clustering is minimum. In practice, this is done by trying out several different values of K. For each K, the clusters are constructed using an iterative refinement technique in which we begin with an initial set of K clusters; move each pattern to a cluster with which it has the minimum squared distance; and recompute cluster centers. The last two steps are iterated until no pattern is moved from its current cluster. One pass of the algorithm is given in Program 7.1.

One pass of the cluster improvement algorithm takes $O(NMK)$ time on a uniprocessor computer. Generally several passes are needed before an acceptable K and corresponding clustering is obtained. In this chapter we consider clustering on both a hypercube with NM processors as well as on a 64 processor NCUBE hypercube.

Step 1: [Cluster Reassignment]

NewCluster $[i] := q$ such that $d2[i, q] = \min_{0 \le k < K} \{d2[i, k]\}$

{ties are broken arbitrarily}

Step 2: [Termination criterion and cluster update]

if *NewCluster* $[i] = C[i], 0 \le i < N$ **then** terminate

else $C[i] :=$ *NewCluster* $[i], 0 \le i < N$;

Step 3: [Cluster center update]

Recompute *center* $[i, j], 0 \le i < K, 0 \le j < M$ using the new cluster assignments;

Program 7.1 One pass of the iterative cluster improvement algorithm

7.2 *NM* Processor Algorithms

7.2.1 Preliminaries

We shall assume that N, M, and K are powers of 2 and that the number of processors in the hypercube is NM. We shall view these processors as an $N \times M$ array. Further, in this section we shall explicitly consider only the case of an SIMD hypercube. The algorithm for an MIMD hypercube is quite similar.

The initial configuration has $F[i, j]$ in the F register of PE(i, j). That is, $F(i, j) = F[i, j], 0 \le i < N, 0 \le j < M$. Also, the center matrix is stored in the top K rows of the $N \times M$ hypercube such that $center(i, j) = center[i, j], 0 \le i < K, 0 \le j < M$.

In addition to some of the operations developed in Chapter 2 we shall need procedures for *term computation*, *distance computation*, and *summing random access write* (SRAW). These are described next.

Term computation is done independently and in parallel on all $K \times 1$ column windows of the NM PE hypercube. The i'th PE of each such window has an F and *center* value, $F(i)$ and $center(i)$, initially. Each PE, i, of the window computes the K values

$$term[k](i) = (F(i) - center(k))^2, \ 0 \le k < K$$

This computation is done by circulating the *center* values through the $K \times 1$ window as in Program 7.2. The complexity of *SIMDTerm* is $O(K)$.

```
procedure SIMDTerm (X, K);
begin
  {index p of processor at (i, j) is p = iK + j}
  S(p) := center(i, j);
  in(p) := i mod K;
  term[in(p)](p) := (F(p) − S(p))²;
  for q := 1 to K−1 do
  begin
    l := f(log₂M, q);
    S(p) ← S(p^(l));
    in(p) := in(p) ⊕ 2^l;
    term[in(p)](p) := (F(p) − S(p))²;
  end;
end; {of SIMDTerm}
```

Program 7.2 Term computation

Distance computation is done independently and in parallel in all $S \times S$ windows where S is a parameter to the operation. The PE in position (i, j) of the window computes

$$dist(i, j) = \sum_{q=0}^{S-1} (F(i, q) - center(j, q))^2, \ 0 \le i < S, \ 0 \le j < S$$

The computation of *dist* is quite similar to computation of the matrix product $C = A*B$ where C, A, and B are $S \times S$ matrices. In fact, if we let

$$A = F \text{ and } B = center^T \quad \text{(i.e., Transpose of } center\text{)}$$

and replace $A[i, k] * B[k, j]$ by $(A[i, k] - B[k, j])^2$ in the definition of matrix product, we end up with the definition of *dist*. Hence, using S^2 processors, *dist* may be computed as follows:

Step 1: Compute B = Transpose of the *center* matrix

Step 2: Use the matrix product procedure *MultSquare* of Chapter 3 to "multiply" F and B. However, each time two terms of F and B are to be multiplied, compute the square of their difference instead.

An $S \times S$ matrix stored one element per processor can be transposed in $O(\log S)$ time as transpose is a BPC permutation. Further, two such matrices can be "multiplied" in $O(S)$ time using S^2 processors and a modified version of procedure *MultSquare*. Hence the complexity of the resulting distance computation procedure is $O(S)$.

An SRAW is a version of the combining random access write of Chapter 2 in which the elements being combined are summed. This is done in $K \times 1$ column windows. The K PEs of a window originate data $A(i)$ that is to be sent to the $dest(i)$'th PE in the window. If two or more PEs have data that is to be sent to the same PE, then their sum is needed at the destination PE. Thus, following the operation, the j'th PE in the $K \times 1$ window has

$$B(j) = \sum_{dest(i)=j} A(i), \; 0 \le j < K$$

This can be done in $O(\log^2 K)$ time using the combining random access write algorithm of Chapter 2. In this, when two A's reach the same PE, they are replaced by a single A which is the sum of the two.

7.2.2 Cluster Reassignment

To parallelize Program 7.1 we need to parallelize the cluster reassignment and cluster update steps (i.e., steps 1 and 3). For cluster reassignment we shall consider the two cases:

(1) Each processor has $O(K)$ memory

(2) Each processor has $O(1)$ memory

Since cluster updating does not benefit from more than $O(1)$ memory per processor, we consider only one case for this step.

7.2.2.1 $O(K)$ Memory Per Processor

The cases $K \leq M$ and $K > M$ result in two slightly different algorithms. The algorithm for the case $K \leq M$ is given in Program 7.3 while that for the case $K > M$ is given in Program 7.4. In Program 7.3, we begin by broadcasting the $K \times M$ cluster center matrix to the remaining $N/K - 1$ $K \times M$ windows of the $N \times M$ hypercube. This is done using a window broadcast. Next, in Step 2, PE(i, j) computes $term[q]$ = $(F[i, j] - center[q, j])^2$, $0 \leq q < K$. This is done by circulating the center values through column windows of size $K \times 1$. The objective of Steps 3 and 4 is to compute $d2(i, k)$, $0 \leq i < K$. $d2(i, k)$ is stored in the $d2$ register of $PE(i, k)$. First, in Step 3 the j'th PE in each $1 \times K$ row window computes the sum of the K $term[j]$ values in the window (i.e., $A(j)$ = $\sum_{q \in 1 \times K\ window} term[j](q)$, $0 \leq j < K$ is computed in all $1 \times K$ windows). This is done using consecutive sum in $1 \times K$ windows. Next, the PEs in the first $1 \times K$ window of each row sum up the values computed by the corresponding PEs in the $1 \times K$ windows in their row. This gives the K $d2$ values for the pattern represented in the row. The minimum of these can be found using a data sum with add replaced by *min*. Once the new cluster for each pattern is known, it can be broadcast to all the PEs in the pattern row (Step 6) for later use.

Step 1: Broadcast the $K \times M$ cluster center matrix in the top $K \times M$ window to the remaining $N/K - 1$ $K \times M$ windows.

Step 2: The PEs in each $K \times 1$ column window compute $term[q]$, $0 \le q < K$.

Step 3: The PEs in each $1 \times K$ row window compute, in A, the consecutive sum of *term* (Chapter 2).

Step 4: The values of A are summed up over the M/K $1 \times K$ windows in each row. This results in $d2(i, k) = \sum_{j=0}^{M-1} term[k](i, j)$, $0 \le i < N$, $0 \le k < K$.

Step 5: Compute $NewCluster(i, 0) = NewCluster[i]$ by finding q such that $d2(i, q) = \min_{0 \le k < K} \{ d2(i, k) \}$.

Step 6: Broadcast $NewCluster(i, 0)$ to the remaining $M-1$ PEs in the i'th row, $0 \le i < N$.

Program 7.3 O(K) memory cluster assignment $K \le M$ (Ranka and Sahni 1988d)

Analyzing the complexity of Program 7.3, we see that step 1 takes $O(\log(N/K))$ time; steps 2 and 3 each take $O(K)$ time; step 4 takes $O(\log(M/K))$ time; step 5 takes $O(\log K)$ time; and step 6 takes $O(\log M)$ time. So the overall complexity of Program 7.3 is $O(K + \log NMK))$.

The algorithm for the case $K > M$ is given in Program 7.4. The strategy is similar to that for the case $K \le M$ and the asymptotic complexity of Program 7.4 is the same as that for Program 7.3.

Step 1: Broadcast the $K \times M$ cluster center matrix in the top $K \times M$ window to the remaining $N/K - 1$ $K{\times}M$ windows.

Step 2: The PEs in each $K \times 1$ column window compute $term[q]$, $0 \le q < K$.

Step 3: Each row forms a $1{\times}M$ window for the consecutive sum operation. This operation is to be repeated K/M times. On the i'th iteration $term[iM + j]$, $0 \le j < M$ of each PE are involved in the operation. Thus each PE computes K/M A values, $A[0..K/M - 1]$.

Step 4: At this time, the PEs in each row have K A values. Each represents a different $d2(i, k)$ value. Each PE computes $D = \min_{0 \le j < M/K} \{ A[j] \}$.

Step 5: PE(i, 0) computes $NewCluster[i]$ by computing the minimum D in its row and the cluster index corresponding to this.

Step 6: Broadcast $NewCluster(i, 0)$ to the remaining $M-1$ PEs in the i'th row, $0 \le i < N$.

Program 7.4 O(K) memory cluster assignment $K > M$ (Ranka and Sahni 1988d)

7.2.2.2 O(1) Memory Per Processor

Once again, we need to develop different algorithms for the two cases $K \le M$ and $K > M$. Program 7.5 gives the algorithm for the former case and Program 7.6 for the latter. The complexity of each algorithm is O(K + log(NMK)).

Let us go through the steps of Program 7.5. Recall that this algorithm is for the case $K \le M$. First, the cluster center window is broadcast such that it resides in all $K{\times}M$ windows of the NM PE hypercube. The objective of Steps 2 and 3 is to compute in PE(i, j), $d2(i, j) = \sum_{q=0}^{M-1} (F(i, q) - center(j, q))^2$, $0 \le i < N$, $0 \le j < K$. In Step 2, PE(i, j)

Step 1: Broadcast the $K \times M$ cluster center matrix in the top $K \times M$ window to the remaining $N/K - 1$ $K \times M$ windows.

Step 2: The PEs in each $K \times K$ window perform a distance calculation. The result is left in the *dist* registers of the PEs.

Step 3: The *dist* values in the M/K $1 \times K$ windows of each row are summed using a window sum. The result of this is left in the $d2$ registers of the PEs in the first $1 \times K$ window of each row.

Step 4: Compute $NewCluster\,(i, 0) = NewCluster\,[i]$ by finding q such that

$$d2(i, q) = \min_{0 \le k < K} \{ d2(i, k) \}.$$

Step 5: Broadcast $NewCluster\,(i, 0)$ to the remaining $M-1$ PEs in the i'th row, $0 \le i < N$.

Program 7.5 O(1) memory cluster assignment $K \le M$ (Ranka and Sahni 1988d)

computes in $dist\,(i, j)$ the sum

$$dist\,(i, j) = \sum_{r=0}^{K-1} (F\,(i, lK + r) - center\,(j, l + r))^2, \; 0 \le i < N, \; 0 \le j \le M$$

where $l = \lfloor j/K \rfloor$. Then, in Step 4, $d2$ is computed by adding the *dist* values in corresponding PEs of the $1 \times K$ windows of the $1 \times M$ rows. Once $d2$ has been computed, the new cluster values are easily obtained and broadcast to all PEs representing the pattern.

Step 1: Broadcast the $K \times M$ cluster center matrix in the top $K \times M$ window to the remaining $N/K - 1$ $K \times M$ windows.

Step 2: In each $K \times M$ window regard the $K \times M$ cluster center matrix as K/M $M \times M$ cluster center matrices. These will be circulated through the K/M $M \times M$ windows of the larger $K \times M$ window. As a result, each $M \times M$ cluster center window will visit each $M \times M$ PE window exactly once. Whenever a new $M \times M$ cluster center window is receieved, the $M \times M$ PE window does Steps 3 and 4. I.e., these are done a total of K/M times.

Step 3: Each $M \times M$ window does a distance computation. Because of the window size used, each computed distance represents the squared distance between a pattern and a cluster.

Step 4: Each PE remembers the smallest distance value it has computed so far. It also remembers the cluster center index that corresponds to this.

Step 5: Compute $NewCluster(i, 0)$ by finding q such that

$$d2(i, q) = \min_{0 \le k < M} \{ d2(i, k) \}$$

and using the cluster center index remembered by PE(i, q)

Step 6: Broadcast $NewCluster(i, 0)$ to the remaining $M-1$ PEs in the i'th row, $0 \le i < N$.

Program 7.6 O(1) memory cluster assignment $K > M$ (Ranka and Sahni 1988d)

7.2.3 Cluster Update

For this operation, we assume that PE(i, j) in the top most $K \times M$ window has values $FeatureSum(i, j)$ and $Number(i, j)$ defined as:

$$FeatureSum(i, j) = \sum_{q \in S_i} F(q, j), 0 \le i \le K, 0 \le j < M$$

$$Number(i, j) = |S_i|, 0 \le i < K, 0 \le j < M$$

The algorithm to update the cluster centers is given in Program 7.7. Steps 1 and 2 are performed in $K \times M$ windows. The (i, j) PE in each such window computes the change in $FeatureSum(i, j)$ and $Number(i, j)$ contributed by the patterns in this window. These two steps can be restricted to PEs for which $NewCluster(i, j) \neq C(i, j)$. In steps 3 and 4 the topmost window accumulates the sum of these changes. Steps 5 through 8 update the clustering data.

For the complexity analysis, we see that Steps 1 and 2 each take $O(\log^2 K)$ time while Steps 3 and 4 each take $O(\log(N/K))$ time. Steps 5 through 8 take O(1) time. The asymptotic complexity of Program 7.7 is therefore $O(\log^2 K + log(N/K))$.

Combining this with the complexity of our procedures for cluster reassignment we get $O(K + \log(NMK))$ as the complexity of our algorithms for one pass of Program 7.1. This is the case regardless of whether the amount of memory available is $O(K)$ or $O(1)$. The complexity may be reduced to $O(\log(NMK))$ per pass if NMK processors are available (Ranka and Sahni 1988d).

7.3 Clustering On An NCUBE Hypercube

Consider the following two cases for the clustering problem on an NCUBE hypercube:

1. Host-to-host: The pattern and cluster information is initially in the host and the result is to be left in the host also.
2. Hypercube-to-hypercube: The pattern and cluster information is initially in the hypercube processors and the result is to be left here.

Let P be the number of hypercube processors. Assume that the N feature vectors that constitute the feature matrix are distributed equally among the P processors and that the center matrix is located initially at processor 0. Also assume that each processor has enough memory to hold

---

Step 1: PE(i, j) does an SRAW of $F(i, j)$ to the $(NewCluster(i, j), j)$ PE in its $K \times M$ window. It also does an SRAW of $-F(i, j)$ to the $(C(i, j), j)$ PE in its $K \times M$ window. Note that both SRAWs involve data movement in $K \times 1$ column windows only. Let the resulting sum in PE(i, j) be $A(i, j)$.

Step 2: PE(i, j) does an SRAW of $+1$ to the $(NewCluster(i, j), j)$ PE in its $K \times M$ window. It also does an SRAW of -1 to the $(C(i, j), j)$ PE in its $K \times M$ window. Once again both SRAWs involve data movement in $K \times 1$ column windows only. Let the resulting sum in PE(i, j) be $B(i, j)$.

Step 3: The A values of corresponding PEs in the N/K $K \times M$ windows are added using window sum. The results are in the D registers of the topmost $K \times M$ window.

Step 4: The B values of corresponding PEs in the N/K $K \times M$ windows are added using window sum. The results are in the E registers of the topmost $K \times M$ window.

Step 5: $FeatureSum(i, j) := FeatureSum(i, j) + D(i, j), 0 \le i < K, 0 \le j < M$

Step 6: $Number(i, j) := \min\{\infty, Number(i, j) + E(i, j)\}, 0 \le i < K, 0 \le j < M$

Step 7: $center(i, j) := FeatureSum(i, j) / Number(i, j), 0 \le i < K, 0 \le j < M$

Step 8: $C(i, j) := NewCluster(i, j), 0 \le i < K, 0 \le j < M$

Program 7.7 Cluster updating (Ranka and Sahni 1988d)

its share of the pattern feature matrix and the whole cluster center matrix. Program 7.8 gives the clustering algorithm for the host-to-host case. The algorithm is self explanatory. Steps 1 and 8 are to be omitted in the hypercube-to-hypercube case.

The NCUBE/7 run times for P = 1, 2, 4, 8, 16, 32, 64; N = 512, 1024; M = 20; and K = 16, 32, 64 for 10 iterations are given in Figure 7.1 for the host-to-host case and in Figure 7.2 for the hypercube-to-hypercube case.

The speedup and efficiency for $N = 512$, $K = 32$, and $M = 20$ for 10 iterations are given in Figure 7.3. Figure 7.4 gives these figures for the case $N = 1024$, $K = 32$, $M = 20$, and number of iterations = 10. From Figure 7.3, we see that when $N = 512$, we get greater then 80% efficiency so long as the number of processors is no more than 16. When $P = 64$, the efficiency drops to approximately 50%. As one would expect, when N is increased the efficiency will also increase. With an N of 1024, the efficiency for $P = 64$ is approximately 60% (since the available memory on a single processor of the NCUBE is insufficient to solve an $N = 1024$ instance on one processor, the $P = 1$ time is estimated from the $P = 2$ time using an efficiency of 0.994. This is the efficiency for the case $N = 512$).

Step 1: Receive partial pattern matrix from host. Processor s receives $F[i, j]$, $s(N/P) \le i < (s+1)(N/P)$, $0 \le s < P$. Processor 0 also receives the cluster center matrix.

Step 2: Steps 3 through 7 are repeated *iteration* number of times.

Step 3: Processor 0 broadcasts the cluster center matrix to all other processors.

Step 4: Each processor calculates the new clusters for each pattern using the cluster center matrix.

Step 5: Each processor s calculates

$$T[s][i, j] = \sum_{a \in S_i} F[a, j], \quad s(N/P) \le i < (s+1)(N/P), 0 \le j < M$$

$$N[s][i] = \sum_{a \in S_i} 1, \quad s(N/P) \le i < (s+1)(N/P)$$

where S_i denotes the new i'th cluster.

Step 6: At processor 0, the following information is gathered

$$X[i, j] = \sum_{s=0}^{P-1} T[s][i, j], \quad 0 \le i < K, 0 \le j < M$$

$$Y[i] = \sum_{s=0}^{P-1} N[s][i], \quad 0 \le i < K$$

This is done using a binary tree scheme. At each stage the processor receiving the information adds its information to the received information and sends it to its parent.

Step 7: Processor 0 calculates the new cluster center matrix.

Step 8: Each processor sends the information about the final value of S_i, $(s(N/P) \le i < (s+1)(N/P))$ to the host.

Program 7.8 Host-to-host clustering algorithm for the NCUBE (Ranka and Sahni 1988d)

	# of clusters		
# of processors	16	32	64
1	44.188	86.842	171.481
2	22.346	43.659	86.286
4	11.463	22.282	43.921
8	6.080	11.700	22.941
16	3.440	6.504	12.632
32	2.184	4.015	7.675
64	1.632	2.889	5.405

(a) 512 patterns

	# of clusters		
# of processors	16	32	64
2	44.519	86.993	171.942
4	22.651	44.053	86.858
8	11.778	22.689	44.510
16	6.390	12.099	23.518
32	3.760	6.912	13.218
64	2.520	4.439	8.278

(b) 1024 patterns
Times are in seconds
Number of features = 20
Number of iterations = 10

Figure 7.1 Host-to-host times

	# of clusters		
# of processors	16	32	64
1	44.182	86.568	171.341
2	22.228	43.544	86.175
4	11.303	22.128	43.779
8	5.891	11.516	22.768
16	3.236	6.307	12.450
32	1.959	3.799	7.477
64	1.372	2.641	5.178

(a) 512 patterns

	# of clusters		
# of processors	16	32	64
2	44.284	86.760	171.714
4	22.331	43.736	86.549
8	11.405	22.321	44.153
16	5.993	11.709	23.142
32	3.338	6.500	12.823
64	2.062	3.991	7.851

(b) 1024 patterns
Times are in seconds
Number of features = 20
Number of iterations = 10

Figure 7.2 Hypercube-to-hypercube times

	P	1	2	4	8	16	32	64
Host-to-host	Speedup	1.000	1.989	3.897	7.422	13.352	21.629	30.059
	Efficiency	1.000	0.994	0.974	0.928	0.834	0.676	0.470
Hypercube-to-hypercube	Speedup	1.000	1.988	3.910	7.517	13.725	22.787	32.778
	Efficiency	1.000	0.994	0.978	0.939	0.857	0.712	0.512

Number of clusters = 32
Number of features = 20
Number of iterations = 10

Figure 7.3 512 patterns: speedup and efficiency

	P	1	2	4	8	16	32	64
Host-to-host	Speedup	1.000	1.988	3.926	7.622	14.294	25.020	38.959
	Efficiency	1.000	0.994	0.981	0.952	0.893	0.782	0.609
Hypercube-to-hypercube	Speedup	1.000	1.988	3.944	7.727	14.730	26.535	43.217
	Efficiency	1.000	0.994	0.986	0.966	0.921	0.829	0.675

Number of clusters = 32
Number of features = 20
Number of iterations = 10

Figure 7.4 1024 patterns: speedup and efficiency (based on estimated time for 1 processor)

Chapter 8

Image Transformations

8.1 Introduction

In this chapter we develop hypercube algorithms for shrinking, expanding, translation, rotation, and scaling of an $N \times N$ image. We assume that N is a power of 2 and that N^2 processors are available. These are viewed as forming an $N \times N$ array. Row major indexing is used for SIMD hypercubes and the Gray code scheme for MIMD hypercubes. The image pixel $I[i, j]$ is mapped to the hypercube processor in position (i, j), $0 \leq i, j < N$, of the array view of the hypercube.

We shall use slightly modified forms of two of the operations developed in Chapter 2. These are:

(1) *End off shift*

The shift operation of Chapter 2 is a wraparound shift in that data from one end of the window wraps around to the other end during the shift. The wraparound feature of this shift operation is easily replaced by an end off *zero* fill feature. In this case, $A(qW + j)$ is replaced by $A(qW + j - i)$, so long as $0 \leq j - i < W$ and by 0 otherwise. This change does not affect the asymptotic complexity of the shift operation. Let *SIMDEShift* (A, i, W) and *MIMDEShift* (A, i, W), denote the end off shift procedures for SIMD and MIMD hypercubes, respectively.

(2) *Row and column reordering*
These are special cases of the combining random access write (RAW) operation. In a row reordering the destination processor for data in any PE is another PE in the same row. Hence, it is sufficient for each PE, p, to simply have a value $dest(p)$ which gives the column index of the destination PE. Furthermore, the *dest* values in each row of the $N \times N$ processor array are either nondecreasing left to right for all rows or nonincreasing left to right for all rows. Because of this monotonicity of the *dest* values, the sort step of the RAW algorithm may be replaced by a step that ranks and concentrates the data.

When *dest* is a nonincreasing function the ranking step does a reverse ranking (i.e., right to left rather than left to right). Alternatively, we can reverse the row in logarithmic time using the BPC algorithm of Chapter 2.

In case $dest(P)$ is not in the range $[0, N-1]$, the data from processor P is not routed anywhere. Since the modified RAWs can be done on all rows in parallel, the time required for row reordering is $O(d \log N)$ where d is the maximum number of processors in any row that have the same *dest*. In case the combining function is an associative operation like min, max, or add the time can be reduced to $O(\log N)$.

Column reordering is the column analog of row reordering. It is performed in an analogous manner.

8.2 Shrinking And Expanding

8.2.1 Problem Formulation

The *neighborhood* of the image point $[i, j]$ is defined to be the set

$$nbd(i, j) = \{ [u, v] \mid 0 \le u < N, 0 \le v < N, \max\{|u-i|, |v-i|\} \le 1\}$$

The *q-step shrinking* of I is defined in (Rosenfeld and Kak 1982 and Rosenfeld 1987) to be the $N \times N$ image S^q such that

$$S^q[i, j] = \min_{[u, v] \in nbd(i, j)} \{I[u, v]\}, q = 1, 0 \le i < N, 0 \le j < N$$

and

$$S^q[i, j] = \min_{[u, v] \in nbd(i, j)} \{S^{q-1}[u, v]\}, q > 1, 0 \le i < N, 0 \le j < N$$

Similarly, the *q-step expansion* of I is defined to be an $N \times N$ image E^q such that

$$E^q[i, j] = \max_{[u, v] \in nbd(i, j)} \{I[u, v]\}, q = 1, 0 \le i < N, 0 \le j < N$$

and

$$E^q[i, j] = \max_{[u, v] \in nbd(i, j)} \{E^{q-1}[u, v]\}, q > 1, 0 \le i < N, 0 \le j < N$$

When the images are binary, the min and max operators in the above definitions may be replaced by *and* and *or* respectively. Let $B_{2q+1}[i, j]$ denote the block of pixels

$$\{[u, v] \mid 0 \le u < N, 0 \le v < N, \max\{|u-i|, |v-j|\} \le q\}$$

Then $nbd(i, j) = B_3[i, j]$. In (Rosenfeld 1987), it is shown that

$$S^q[i, j] = \min_{[u, v] \in B_{2q+1}(i, j)} \{I[u, v]\}, \quad 0 \le i < N, \ 0 \le j < N \tag{8.1}$$

and

$$E^q[i, j] = \max_{[u, v] \in B_{2q+1}(i, j)} \{I[u, v]\}, \quad 0 \le i < N, \ 0 \le j < N$$

Our remaining discussion of shrinking and expanding will explicitly consider shrinking only. Our algorithms for shrinking can be easily transformed to expanding algorithms of the same complexity. This transformation simply requires the replacement of every min by a max and a change in the end off shift fill in from ∞ to $-\infty$. In the case of binary images the min and max operators may be replaced by *and* and *or* respectively and the end off shift fill in of ∞ and $-\infty$ by 1 and 0, respectively.

Let $R^q[i, j]$ be defined as below

$$R^q[i, j] = \min_{[i, v] \in B_{2q+1}(i, j)} \{I[i, v]\}, \quad 0 \le i < N, \ 0 \le j < N \tag{8.2}$$

From (8.1), it follows that

$$S^q[i, j] = \min_{[u, j] \in B_{2q+1}(i, j)} \{R^q[u, j]\}, \quad 0 \le i < N, \ 0 \le j < N \tag{8.3}$$

When an $N \times N$ image is mapped onto an $N \times N$ MIMD or SIMD hypercube using the mappings of Chapter 1, the rows and columns of the mappings are symmetric. Consequently, the algorithms to compute R^q and S^q from (8.2) and (8.3) are very similar. Hence, in the sequel we consider the computation of R^q only. We assume $q = 2^k$.

8.2.2 MIMD Shrinking

On an MIMD hypercube R^q for $q = 2^k$ may be computed using the algorithm of Program 8.1. The computation of R is done in two stages. These are obtained by decomposing (8.2) into

$$left^q[i, j] = \min_{\substack{[i, v] \in B_{2q+1}[i, j] \\ v \le i}} \{I[i, v]\}, \ 0 \le i < N, \ 0 \le j < N$$

$$right^q[i, j] = \min_{\substack{[i, v] \in B_{2q+1}[i, j] \\ v \ge i}} \{I[i, v]\}, \ 0 \le i < N, \ 0 \le j < N$$

$$R^q[i, j] = \min \{ \ left^q[i, j], right^q[i, j] \ \} \ \ 0 \le i < N, \ 0 \le j < N$$

One may verify that following the first **for** loop iteration with $i = a$, $left(p)$ is the min of the pixel values in the left 2^a processors and that in its own I register, $0 \le a < k$. To complete the computation of $left(p)$ we need also to consider the pixel value 2^k units to the left and on the same image row. This is done by a rightward shift of 2^k. The shift is done by rows (i.e., blocks of size N) with a fill in of ∞. A similar argument establishes the correctness of the second stage computation of *right*.

Since a power of 2 MIMD shift takes O(1) time, it follows that the complexity of procedure *MIMDShrink* is $O(k)$. Once R^q, $q = 2^k$, has been computed, S^q may be computed using a similar algorithm.

8.2.3 SIMD Shrinking

Since a shift of 2^i in a window of size N takes $O(\log(N/2^i))$ time on an SIMD hypercube, a simple adaptation of *MIMDShrink* to SIMD hypercubes will result in an algorithm whose complexity is $O(k\log N)$. We can do better than this using a different strategy.

```
procedure MIMDShrink;
{Compute R^q for q = 2^k on an MIMD hypercube}
begin
  {compute min of the left 2^k pixels on the same row}
  {MIMDEShift does an ∞ fill instead of a 0 fill}
  left(p) := I(p);
  for i :=0 to k - 1 do
  begin
    C(p) := left(p);
    MIMDEShift(C, 2^i, N);
    left(p) := min {left(p), C(p)};
  end
  C(p) := I(p);
  MIMDEShift(C, 2^k, N);
  left(p) := min {left(p), C(p)};

  {compute min of the right 2^k pixels on the same row}
  right(p) := I(p);
  for i :=0 to k - 1 do
  begin
    C(p) := right(p);
    MIMDEShift(C,-2^i, N);
    right(p) := min {right(p), C(p)};
  end
  C(p) := I(p);
  MIMDEShift(C, -2^k, N);
  right(p) := min {right(p), C(p)};

  R(p) := min {left(p), right(p)}
end;
```

Program 8.1 MIMD computation of R^q (Ranka and Sahni 1989b)

The $N \times N$ image is mapped onto the $N \times N$ hypercube using the row major mapping. R^q for $q = 2^k$ may be computed by considering the N processors that represent a row of the image as comprised of several blocks of size 2^k each (see Figure 8.1).

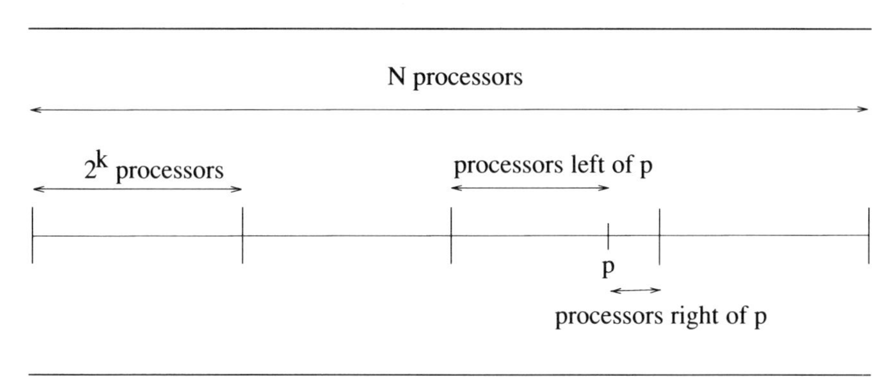

Figure 8.1 2^k blocks of processors

Each processor p computes

left(p) = minimum of pixel values to the left of p but within the same 2^k block

right(p) = minimum of pixel values to the right of p but within the same 2^k block

Now, $R^q(p)$ is the minimum of:

(a) I(p)
(b) *left*(p)
(c) *right*(p)
(d) *left*($p + q$) provided $p + q$ is in the same row
(e) *right*($p - q$) provided $p - q$ is in the same row

Note that this is true even if q is not a power of 2 and we use $k = \lfloor \log_2 q \rfloor$ in the definition of *left* and *right*. *left(p)* and *right(p)* for 2^k blocks may be computed by first computing these for 2^0 blocks, then for 2^1 blocks, then 2^2 blocks and so on. Let *whole* (p) be the minimum of all pixels in the block that currently contains PE p. For 2^0 blocks, we have

$$left(p) = right(p) = \infty$$

$$whole(p) = I(p)$$

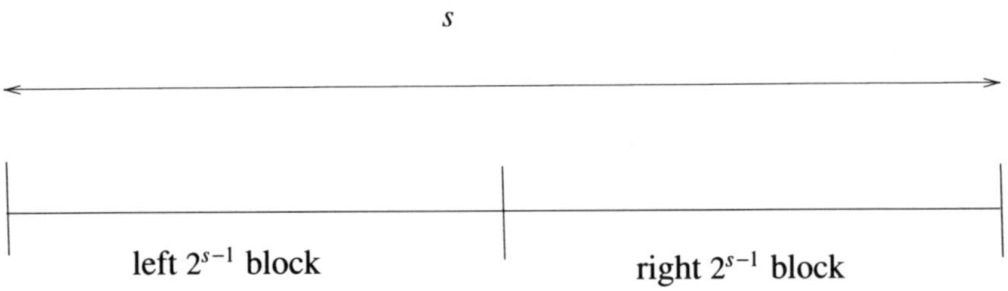

Figure 8.2 A 2^s block of processors

Each 2^s block for $s > 0$ consists of two 2^{s-1} blocks as shown in Figure 8.2. One is the left 2^{s-1} block and the other the right 2^{s-1} block. The PEs in the left 2^{s-1} block have bit $s-1 = 0$ while those in the right one have bit $s-1 = 1$. Let us use a superscript to denote block size. So, $left^s(p)$ denotes *left(p)* when the block size is 2^s. We see that when p is in the left 2^{s-1} block,

$$left^s(p) = left^{s-1}(p)$$

$$right^s(p) = \min\{right^{s-1}(p), whole^{s-1}(p + 2^{s-1})\}$$

$$whole^s(p) = \min\{whole^{s-1}(p), whole^{s-1}(p + 2^{s-1})\}$$

and when p is in the right 2^{s-1} block,

$$left^{s}(p) = \min\{left^{s-1}(p), whole^{s-1}(p - 2^{s-1})\}$$

$$right^{s}(p) = right^{s-1}(p)$$

$$whole^{s}(p) = \min\{whole^{s-1}(p), whole^{s-1}(p - 2^{s-1})\}$$

Program 8.2 implements the strategy just developed. Its complexity is $O(\log N)$. The algorithm can also be used when q is not a power of 2 by simply defining $k = \lfloor \log q \rfloor$. The complexity remains $O(\log N)$.

8.3 Translation

This operation requires moving the pixel at position $[i, j]$ to the position $[i + a, j + b]$, $0 \leq i < N$, $0 \leq j < N$ where a and b are given and assumed to be in the range $0 \leq a, b \leq N$. Translation may call for image wraparound in case $i + a \geq N$ or $j + b \geq N$. Alternatively pixels that get moved to a position $[c, d]$ with either $c \geq N$ or $d \geq N$ are discarded and pixel positions $[i, j]$ with $i < a$ or $j < b$ get filled with zeroes. Regardless of which alternative is used, image translation can be done by first shifting by a along rows (circular shift for wraparound or zero fill right shift for no wraparound) and then shifting by b along columns. Unless a and b are powers of 2, the time complexity is $O(\log N)$ on both an SIMD and an MIMD hypercube. When a and b are powers of 2, the translation takes $O(1)$ time on an MIMD hypercube.

```
procedure SIMDShrink;
{Compute R^q for q = 2^k on an SIMD hypercube}
begin
  {initialize for 2^0 blocks}
  whole(p) := I(p);
  left(p) := ∞;
  right(p) := ∞;

  {compute for 2^(i+1) blocks}
  for i :=0 to k - 1 do
  begin
    C(p) := whole(p);
    C(p) ← C(p^(i));
    left(p) := min {left(p), C(p)}; (p^(i) = 1)
    right(p) := min {right(p), C(p)}; (p^(i) = 0)
    whole(p) := min {whole(p), C(p)};
  end

  R(p) :=min { I(p), left(p), right(p)}
  SIMDEShift(left, -q, N);
  SIMDEShift(right, q, N);
  R(p) :=min { R(p), left(p), right(p)}
end; {of SIMDShrink}
```

Program 8.2 SIMD computation of R^q (Ranka and Sahni 1989b)

8.4 Rotation

The image I is to be rotated θ degrees about the point $[a, b]$ where a and b are integers in the range $[0, N-1]$. Following the rotation, *pixel* $[i, j]$ of I will be at position $[i', j']$ where i' and j' are given by Reeves and Francfort (1985)

$$i' = \lceil (i-a)\cos\theta - (j-b)\sin\theta + a \rceil$$

$$j' = \lceil (i-a)\sin\theta + (j-b)\cos\theta + b \rceil$$

The equations for i' and j', may be simplified to

$$i' = \lceil i\cos\theta - j\sin\theta + A \rceil \qquad (8.4)$$

and

$$j' = \lceil i\sin\theta + j\cos\theta + B \rceil$$

where $A = a(1-\cos\theta) + b\sin\theta$ and $B = b(1-\cos\theta) - a\sin\theta$.

The hypercube rotation algorithm considers several cases for θ. The steps for each of these cases are developed in the following subsections.

8.4.1 $\theta = 180^0$

In this case, $i' = -i + a$, and $j' = -j + b$. The rotation can be performed as follows:

Step 1: [Column reordering]
Each processor, p, sets $dest(p) = -i + a$ where i is the row number of the processor. Next, a column reordering is done.

Step 2: [Row reordering]
Each processor, p, sets $dest(p) = -j + b$ where j is the column number of the processor and a row reordering is done.

Note that the dest values in each column in Step 1 and those in each row in Step 2 are in decreasing order. Step 1 sends all pixels to their correct destination row while Step 2 sends them to the correct column. Since the *dest* values are distinct the row and column reordering of steps 1 and 2 can be also be done by a reversal followed by a shift. While this is simpler than using the modified RAW scheme discussed earlier, its asymptotic complexity is the same. The complexity of the two step 180^0 rotation algorithm is $O(\log N)$.

8.4.2 $\theta = \pm 90^0$

The case $\theta = 90^0$ and $\theta = -90^0$ are quite similar. We explicitly consider only the case $\theta = 90^0$. Now, $i' = -j + a + b$ and $j' = i - a + b$.

The steps in the rotation are:

Step 1: [Transpose]
Transpose the image so that $I^{new}[i, j] = I^{old}[j, i]$

Step 2: [Column Reorder]
Each processor sets $dest(p) = a + b - i$ where i is the row number of the processor. Next, a column reordering is done.

Step 3: [Shift]
A rightward shift of $-a + b$ is performed on each row of the image.

In a 90^0 rotation the pixel originally at $[i, j]$ is to be routed to $[-j + a + b, i - a + b]$. Step 1 routes the pixel to position $[j, i]$; Step 2 routes it to $[a + b - j, i]$; and Step 3 to $[a + b - j, i - a + b]$. The transpose of Step 1 can be performed in $O(\log N)$ time using the BPC algorithm of Chapter 2. The overall complexity is $O(\log N)$. Once again, the column reordering of Step 2 can be done by a column reversal followed by a shift. This does not change the asymptotic complexity.

8.4.3 $|\theta| \le 45^0$

We explicitly consider the case $0 \le \theta \le 45^0$ only. The case $-45^0 \le \theta < 0$ is similar. The steps for the case $0 \le \theta \le 45^0$ are:

Step 1: [Column Reorder]
Set $dest(p) = \lceil i\cos\theta - j\sin\theta + A \rceil$ where i is the row number and j the column number of processor p. Since j is the same in a column, $dest(p)$ is nondecreasing in each column. Hence a column reordering can be done. All data with the same destination are routed to that destination.

Step 2: [Row Reorder]
Set $dest(p) = \lceil i\tan\theta + j sec\theta - A\tan\theta + B \rceil$ where i and j are respectively, the row and column numbers of processor p. A row reordering is now performed.

Step 3: [Shift]
Pixels that need to be shifted left by one along rows are shifted.

Step 1 sends each pixel to its correct destination row. Since $0 \le \theta \le 45^0$, $1/\sqrt{2} \le \cos\theta \le 1$. Hence, each processor can have at most 2 pixels directed to it. The column reordering of Step 1 is done such that both these reach their destination. Following this, the pixel(s) in the processor at position $[i, j]$ originated in processors in column j and row

$$\frac{i + j\sin\theta - A - \delta}{\cos\theta}$$

where $0 \le \delta < 1$ accounts for the ceiling function in (4). From (4), it follows that these pixels are to be routed to the processors in row i and column $j = \lceil y \rceil$ where y is given by

$$y = \left(\frac{i + j\sin\theta - A - \delta}{\cos\theta}\right)\sin\theta + j\cos\theta + B$$

$$= i\tan\theta + j\left(\frac{\sin^2\theta + \cos^2\theta}{\cos\theta}\right) - A\tan\theta - \delta\tan\theta + B$$

$$= i\tan\theta + j sec\theta - A\tan\theta + B - \delta\tan\theta$$

In Step 2, the pixels are first routed to the column $\lceil i\tan\theta + j sec\theta - A\tan\theta + B \rceil$. Then, in Step 3, we account for the $\delta\tan\theta$ term in the formula for y. For $0 \le \theta \le 45^0$, $\tan\theta$ is in the range $[0, 1]$. Since $0 \le \delta < 1$, $0 \le \delta\tan\theta < 1$, the pixels need to be shifted leftwards on the rows by at most 1. Note that since $1 \le sec\theta \le \sqrt{2}$ for $0 \le \theta \le 45^0$, $dest(p)$ is different for different processors on the same row. One readily sees that $O(\log N)$ time suffices for the rotation.

8.4.4 $0 \le \theta \le 360^0$

Every θ in the range [0,360] can be cast into one of the forms:

(a) $-45 \le \theta' \le 45$
(b) $\pm 90 + \theta', \quad -45 \le \theta' \le 45$
(c) $\pm 180 + \theta', \quad -45 \le \theta' \le 45$

Case (a) was handled in the last subsection. Cases (b) and (c) can be done in two steps. First a $\pm 90^0$ or a 180^0 rotation is done (note that a $+180^0$ and a -180^0 rotation are identical). Next a θ' rotation is performed. This two step process may introduce some errors because of end off conditions from the first step. These can be eliminated by implementing all rotations as wraparound rotations and then having a final cleanup step to eliminate the undesired wraparound pixels.

8.5 Scaling

Scaling an image by s, $s \ge 0$, around position $[a, b]$ requires moving the pixel at position $[i, j]$ to the position $[i', j']$ such that (Lee, Yalamanchali, and Agarwal 1987):

$$i' = \lceil si + a(1 - s) \rceil$$

$$j' = \lceil sj + b(1 - s) \rceil$$

$$0 \le i, j < N.$$

In case $i' \ge N$ or $j' \ge N$, the pixel is discarded. If two or more pixels get mapped to the same location then we have two cases:

(1) only one of these is to survive. The surviving pixel is obtained by some associative operation such as *max*, *min*, *average* etc.

(2) all pixels are to survive.

When $s > 1$, then in addition to routing each pixel to its destination pixel, it is necessary to reconnect the image boundary and fill in the inside of the objects in the image (Lee, Yalamanchali, and Agarwal 1987). The pixel routing can be done in $O((\log N)/s)$ time when $s < 1$ and all pixels to the same destination are to survive. In all other cases, pixel routing takes $O(\log N)$ time. The routing strategy is to perform a row reordering followed by a column reordering. Reconnecting the boundary and filling can be done in $O(\log N)$ time.

CHAPTER 9

SIMD String Editing

9.1 Introduction

The input to the string editing problem consists of two strings $A = a_1 a_2 a_3 \ldots a_{n-1}$ and $B = b_1 b_2 b_3 \ldots b_{m-1}$; and three cost functions C, D, and I where:

$$C(a_i, b_j) = \text{ cost of changing } a_i \text{ to } b_j$$

$$D(a_i) = \text{ cost of deleting } a_i \text{ from A}$$

$$I(b_i) = \text{ cost of inserting } b_i \text{ into A}$$

Three edit functions: *change*, *delete* and *insert* are available. C, D, and I give the cost of one application of each of these functions. The *cost* of a sequence of edit functions is the sum of the costs of the individual functions in the sequence. In the *string edit* problem, we are required to find a minimum cost editing sequence that transforms string A into string B.

The string edit problem is identical to the weighted Levenshtein distance problem (Liu and Fu 1985). The longest common subsequence problem (Wagner and Fischer 1974) and the time warping distance problem (Liu and Fu 1985) are special cases of the string edit problem.

9.2 Dynamic Programming Formulation

Wagner and Fischer (1974) have proposed a single processor dynamic programming solution for the string editing problem. This formulation is in terms of a function *cost* where $cost(i, j) =$ is the cost of a minimum cost edit sequence to transform $a_1 a_2 \cdots a_i$ into $b_1 b_2 \cdots b_j$

The following recurrence for *cost* is easily obtained:

$$cost(i, j) = \begin{cases} 0 & i = j = 0 \\ cost(i-1, 0) + D(a_i) & i > 0, j = 0 \\ cost(0, j-1) + I(b_j) & i = 0, j > 0 \\ cost'(i, j) & i > 0, j > 0 \end{cases} \tag{9.1}$$

where

$$\begin{aligned} cost'(i, j) = \min\{ & cost(i-1, j) + D(a_i) \\ & cost(i-1, j-1) + C(a_i, b_j) \\ & cost(i, j-1) + I(b_j)\} \end{aligned}$$

Once $cost(i, j), 0 \le i < n, 0 \le j < m$ has been computed a minimum cost edit sequence may be found by a simple backward trace from $cost(n-1, m-1)$. This backward trace is facilitated by recording which of the three options for $i > 0, j > 0$ yielded the minimum for each i and j.

The dependencies in the dynamic programming recurrence (9.1) may be represented by a lattice graph (Figure 9.1). The vertex in position (i, j) of the lattice graph represents entry (i, j) of the cost matrix. Each edge of the lattice graph is assigned a weight equal to the cost of the corresponding edit operation. The weights are obtained as follows:

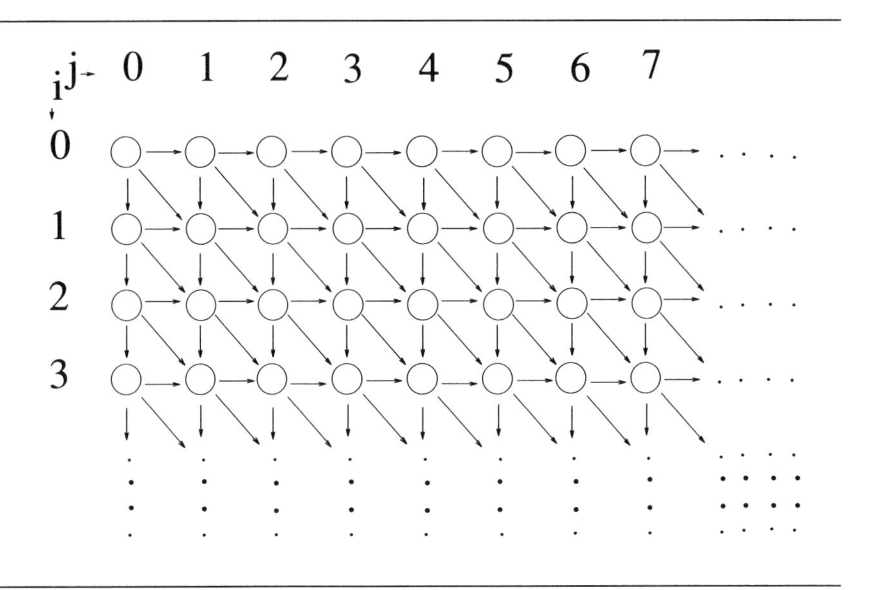

Figure 9.1 Lattice graph

(1) The weight of an edge of type $<(i-1, j), (i, j)>$ is $D(a_i)$

(2) An edge of type $<(i-1, j-1), (i, j)>$ has weight $C(a_i, b_j)$

(3) The weight of an edge of type $<(i, j-1), (i, j)>$ is $I(b_j)$

With this weight assignment, $cost(i, j)$ is the length of a shortest path from vertex $(0, 0)$ to vertex (i, j), $0 \le i < n$, $0 \le j < m$. Using the above dynamic programming recurrence both $cost(n-1, m-1)$ and the actual least cost edit sequence can be found in $O(nm)$ time.

9.3 Shared Memory Parallel Algorithm

9.3.1 Motivation

As noted in the previous section, the string edit problem is really that of finding a shortest path in a lattice graph. A shortest path in an N vertex graph can be found by repeatedly squaring the $N \times N$ cost matrix of the graph $\log_2 N$ times (Dekel, Nassimi, and Sahni 1981). Using the matrix multiplication algorithm of Chapter 2 we can square the cost matrix in $O(\log N)$ time on a hypercube with $N^3/\log N$ processors. So computing the square $\log_2 N$ times will take $O(\log^2 N)$ time. Hence we can compute the shortest path in an N vertex graph in $O(\log^2 N)$ time with $O(N^3/\log N)$ processors. Since the string edit graph (Figure 9.1) has $O(nm)$ vertices, this strategy will result in an $O(\log^2(nm))$ algorithm that requires a hypercube with $O((nm)^3/\log(nm))$ processors. The processor-time product of this algorithm is $O((nm)^3 \log(nm))$. The single processor dynamic programming algorithm has a processor-time product that is only $O(nm)$.

Our concern in this chapter is to develop a fast SIMD hypercube string editing algorithm with a superior processor-time product. We shall do this in two stages (Ranka and Sahni 1988c, Ibarra, Pong, and Sohn 1988). First we formulate (in this section) a parallel algorithm for a shared memory parallel computer. Then, in the next section, we map this parallel algorithm on to an SIMD hypercube. For convenience, we assume $n = m = 2^q$ for some natural number q. The development is easily extended to the case $n \neq m$ and also to the case when n and m are not powers of 2.

Our strategy to find a shortest path from (0, 0) to $(n-1, n-1)$ consists of two phases:

Phase 1: Compute $cost(n-1, n-1)$

Phase 2: Trace back to obtain the path

9.3.2 Computing $cost(n-1, n-1)$

The first phase itself is accomplished in two stages as described below.

Phase 1, Stage 1
The lattice graph is decomposed into $k \times k$ sublattice graphs for some k that is a power of 2. The optimal value of k will be determined later. For each $k \times k$ sublattice graph the shortest distance from each vertex on the top and left boundaries to each vertex on the bottom and right boundaries is found.

Phase 1, Stage 2
The boundary to boundary distances computed in Stage 1 are combined to obtain $cost(n-1, n-1)$.

Figure 9.2 shows a $2a \times 2a$ lattice graph made up of four $a \times a$ lattice graphs. The smaller lattice graphs have been labeled 0 - 3. Let T_i, B_i, L_i, and R_i, respectively, denote the top, bottom, left, and right boundary vertices in the smaller lattice graph i, $0 \le i \le 3$. We shall use the notation $XY(i, j)$ to refer to the shortest distance from the i'th vertex in boundary X to the j'th vertex in boundary Y, $X, Y \in \{T_i, B_i, L_i, R_i\}$. Vertices are numbered $0, ..., a-1$ left to right for top and bottom boundaries and top to bottom for left and right boundaries. Note that vertex 0 of a top boundary is also the vertex 0 of a left boundary. Similarly, vertex $a-1$ of a top boundary is vertex 0 of a right boundary, etc. $T_0R_0(i, j)$ is the length of a shortest path from the i'th vertex of the top boundary of the lattice graph 0 to the j'th vertex of the right boundary of lattice graph 0.

The boundary distances we are to compute for the $2a \times 2a$ lattice graph are:

$$T_0R_1, T_0R_3, T_1R_1, T_1R_3, T_0B_2, T_0B_3, T_1B_2, T_1B_3,$$
$$L_0R_1, L_0R_3, L_2R_1, L_2R_3, L_0B_2, L_0B_3, L_2B_2, L_2B_3$$

Because of the edge structure of our lattice graph (Figure 9.1) we know

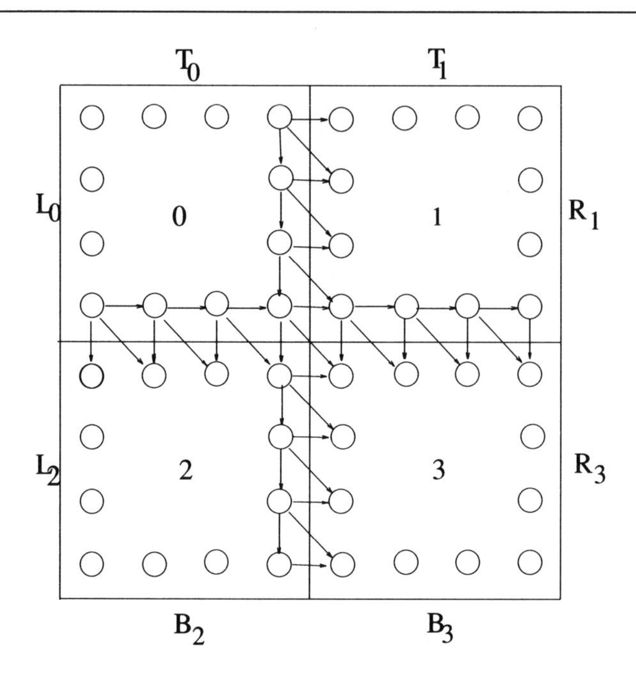

Figure 9.2 An 8×8 lattice graph decomposed into 4 4×4 subgraphs

that all distances in L_2R_1 and T_1B_2 are ∞. From Figure 9.1, we see that $T_0R_1(i, j)$ is given by:

$$T_0R_1(i, j) = \min\{ \min_{0 \le s < a} \{T_0R_0(i, s) + R_0L_1(s, s) + L_1R_1(s, j)\}, \quad (9.2)$$

$$\min_{0 \le s < a-1} \{T_0R_0(i, s) + R_0L_1(s, s+1) + L_1R_1(s+1, j)\}\}$$

Since $R_0L_1(s, s)$ and $R_0L_1(s, s+1)$ are simply edge costs, $T_0R_1(i, j), 0 \le i, j < a$ can be computed if T_0R_0 and L_1R_1 are known. T_0R_0 and L_1R_1 are boundary distances for $a \times a$ lattice subgraphs. The computation of T_0R_3 from $a \times a$ boundary distances is more complex. The equations needed are:

$$T_0R_2(i, j) = \min\{ \min_{0 \le s < a} \{T_0B_0(i, s) + B_0T_2(s, s) + T_2R_2(s, j)\}, \quad (9.3)$$

$$\min_{0 \le s < a-1} \{T_0B_0(i, s) + B_0T_2(s, s+1) + T_2R_2(s+1, j)\}\}$$

$$T_0R'_3(i, j) = \min\{ \min_{0 \le s < a} \{T_0R_2(i, s) + R_2L_3(s, s) + L_3R_3(s, j)\}, \quad (9.4)$$

$$\min_{0 \le s < a-1} \{T_0R_2(i, s) + R_2L_3(s, s+1) + L_3R_3(s+1, j)\}\}$$

$$T_0B_1(i, j) = \min\{ \min_{0 \le s < a} \{T_0R_0(i, s) + R_0L_1(s, s) + L_1B_1(s, j)\}, \quad (9.5)$$

$$\min_{0 \le s < a-1} \{T_0R_0(i, s) + R_0L_1(s, s+1) + L_1B_1(s+1, j)\}\}$$

$$T_0R''_3(i, j) = \min\{ \min_{0 \le s < a} \{T_0B_1(i, s) + B_1T_3(s, s) + T_3R_3(s, j)\}, \quad (9.6)$$

$$\min_{0 \le s < a-1} \{T_0B_1(i, s) + B_1T_3(s, s+1) + T_3R_3(s+1, j)\}\}$$

$$T_0R_3(i, j) = \min\{T_0R_3'(i, j), T_0R''_3(i, j), \quad (9.7)$$

$$T_0B_0(i, a-1) + B_0T_3(a-1, 0) + T_3R_3(0, j)\}$$

T_1R_1 and L_2B_2 for the $2a \times 2a$ graph are the same as for the corresponding $a \times a$ graphs. The equations for T_1R_3, T_0B_2, T_1B_2, T_1B_3, L_0R_1, L_2R_3, L_0B_2, and L_0B_3 the equations are similar to those for T_0R_3. For a 1×1 graph,

$$TR(0, 0) = TB(0, 0) = LR(0, 0) = LB(0, 0) = 0$$

Hence the boundary distances for any $k \times k$ lattice subgraph may be computed by computing these distances for 2×2 subgraphs, then for 4×4 subgraphs, then for 8×8 subgraphs, till a $k \times k$ lattice graph is reached.

After the boundary distances for each $k \times k$ subgraph have been computed, we compute for each $k \times k$ subgraph the shortest distance from vertex (0, 0) (of the whole graph) to each of the vertices on the top, bottom, left and right boundaries of the $k \times k$ subgraph. Figure 9.3 shows an $n \times n$ graph and its composite $k \times k$ subgraphs. The figure is for the case $n = 4k$. The $k \times k$ subgraphs are labeled a–p.

(0,0)

a 1	b 2	c 3	d 4
e 2	f 3	g 4	h 5
i 3	j 4	k 5	l 6
m 4	n 5	o 6	p 7

$(n-1, n-1)$

Figure 9.3 An $n \times n$ subgraph and its composite $k \times k$ subgraphs $n = 4k$

The shortest distances from (0, 0) to the boundary vertices of $(k \times k)$ subgraphs is computed in several iterations. In iteration i the distances to the boundary vertices of all subgraphs assigned the number i in Figure 9.3 are computed.

Let $T_iT_i(l, j)$ be the length of the shortest path from the l'th vertex of the top boundary of the $k \times k$ subgraph i to the j'th vertex of its top boundary, $0 \le l \le j < k$. Clearly, $T_iT_i(l, j) = \sum_{r=l}^{j-1} T_iT_i(r, r+1)$ where $T_iT_i(r, r+1)$ is the cost of the directed edge between the top boundary vertices r and $r+1$. Let $L_iL_i(l, j)$ be the length of the shortest path from the l'th vertex of the left boundary of the $k \times k$ subgraph i to the j'th vertex of its left boundary, $0 \le l \le j < k$. We see that $L_iL_i(l, j) = \sum_{r=l}^{j-1} L_iL_i(r, r+1)$ where $L_iL_i(r, r+1)$ is the cost of the directed edge between the left boundary vertices r and $r+1$. Let $T_i(j)$, $B_i(j)$, $L_i(j)$, and $R_i(j)$, respectively, denote the shortest distance from vertex (0, 0) to the j'th vertex of the top, bottom, left, and right boundaries of the $k \times k$ subgraph i. For subgraph a of

Figure 9.3, we have:

$$T_a(j) = T_aT_a(0, j);\ L_a(j) = L_aL_a(0, j);\ R_a(j) = T_aR_a(0, j);\ B_a(j) = T_aB_a(0, j)$$

where T_aR_a and T_aB_a are boundary distances that have already been computed. For subgraphs b and e of Figure 9.3, we have:

$$L_b(j) = \min\{\min_{0 \le s \le j} \{R_a(s) + R_aL_b(s, s) + L_bL_b(s, j)\}, \min_{0 \le s < j} \{R_a(s) + R_aL_b(s, s+1) + L_bL_b(s+1, j)\}\}$$

$$T_b(j) = L_b(0) + T_bT_b(0, j)$$

$$R_b(j) = \min_{0 \le s \le j} \{L_b(s) + L_bR_b(s, j)\}$$

$$B_b(j) = \min_{0 \le s \le j} \{L_b(s) + L_bB_b(s, j)\}$$

$$T_e(j) = \min\{\min_{0 \le s \le j} \{B_a(s) + B_aT_e(s, s) + T_eT_e(s, j)\}, \min_{0 \le s < j} \{B_a(s) + B_aT_e(s, s+1) + T_eT_e(s+1, j)\}\}$$

$$L_e(j) = T_e(0) + L_eL_e(0, j)$$

$$R_e(j) = \min_{0 \le s \le j} \{T_e(s) + T_eR_e(s, j)\}$$

$$B_e(j) = \min_{0 \le s \le j} \{T_e(s) + T_eB_e(s, j)\}$$

where R_aL_b and B_aT_e are edge costs and L_bR_b, L_bB_b, T_eR_e, and T_eB_e are boundary distances for the respective $k \times k$ subgraphs. The last case to consider is that of subgraph f of Figure 9.3. For this, we obtain:

$$T_f(0) = L_f(0) = \min\{B_a(k-1) + B_aT_f(k-1,0), B_b(0) + B_bT_f(0, 0), R_e(0) + R_eT_f(0, 0)\}$$

$$T_f(j) = \min\{T_f(0) + T_fT_f(0, j),\ \min_{0 \le s \le j}\{B_b(s) + B_bT_f(s, s) + T_fT_f(s, j)\},\ \min_{0 \le s < j}\{B_b(s) + B_bT_f(s, s+1) + T_fT_f(s+1, j)\}\} \quad , \ j > 0$$

$$L_f(j) = \min\{L_f(0) + L_fL_f(0, j),\ \min_{0 \le s \le j}\{R_e(s) + R_eL_f(s, s) + L_fL_f(s, j)\},\ \min_{0 \le s < j}\{R_e(s) + R_eL_f(s, s+1) + L_fL_f(s+1, j)\}\} \quad , \ j > 0$$

$$R_f(j) = \min\{\min_{0 \le s \le j}\{T_f(s) + T_fR_f(s, j)\},\ \min_{0 \le s \le j}\{L_f(s) + L_fR_f(s, j)\}\}$$

$$B_f(j) = \min\{\min_{0 \le s \le j}\{T_f(s) + T_fB_f(s, j)\},\ \min_{0 \le s \le j}\{L_f(s) + L_fB_f(s, j)\}\}$$

The $k \times k$ subgraphs of an $n \times n$ lattice graph (Figure 9.3) may be partitioned into four classes:

(1) Top left corner subgraph (subgraph a of Figure 9.3)

(2) Remaining top boundary subgraphs (subgraphs b, c, and d of Figure 9.3)

(3) Remaining left boundary subgraphs (e, i, and m of Figure 9.3)

(4) All other subgraphs

We see that the formulas obtained above for subgraphs a, b, e, and f can be easily adapted to cover all subgraphs. A close examination of the formulas for a, b, e, and f reveals the following computations are not required.

(1) T and L values of top corner subgraph

(2) T values of the remaining top boundary subgraphs

(3) L values of the remaining left boundary subgraphs

(4) B values of the bottom boundary subgraphs

(5) R values of the right boundary subgraphs (excluding $R(k-1)$ of the bottom right corner subgraph).

Note that $cost(n-1, n-1) = R(k-1)$ of the bottom right corner subgraph.

9.3.3 Traceback

The shortest path from (0, 0) to $(n-1, n-1)$ (i.e., the least cost edit sequence) can be obtained in two stages:

Stage1: Each $k \times k$ subgraph determines the vertex (if any) at which this path enters the subgraph and the vertex (if any) from which it leaves the subgraph.

Stage2: The subgraphs that have an entry and exit vertex determine a shortest path in the subgraph from entry to exit.

9.3.3.1 Subgraph Entry/Exit Vertices

These can be determined easily if with each $L_i(j)$, $T_i(j)$, $B_i(j)$, and $R_i(j)$ computed in Section 9.3.2 we record ''how'' the minimum of the quantities on the right hand side of the respective equation was achieved. So, when computing $B_f(j)$ we will also record a value (X, u), $X \in \{L, T\}$, $0 \le u \le j$ such that $B_f(j) = X_f(u) + X_f B_f(u, j)$. Using this information, we begin at $R_z(k-1)$ where z is the bottom right corner subgraph and work our way back to $T_a(0)$ where a is the top left subgraph. For the graph of Figure 9.3, beginning at $R_p(k-1)$ we obtain the entry vertex for subgraph p. From this entry vertex and the recorded information, we obtain the exit vertex from subgraphs k, o, or l that was used to get to the entry vertex of p. Suppose this exit vertex is in subgraph l. From B_l we obtain the entry vertex for l and so on.

Because of the edge structure of the graph, exactly one of the subgraphs with any given numeric label (cf. Figure 9.3) will have an entry and exit vertex. If there are many possible shortest paths, a tie is broken arbitrarily at each step.

9.3.3.2 Shortest Path In A Subgraph

This can be computed if during the computation of boundary distances, we record 'how' each decision is made. Since $O(k^2 \log k)$ decisions are made, this much memory is needed to record the decision information. The actual path computation follows a process similar to that used to compute the entry/exit vertices. Entry/exit points in $k/2 \times k/2$ blocks are found; then in $k/4 \times k/4$ blocks; etc.

9.4 SIMD Hypercube Mapping

9.4.1 $n^2p,\ 1 \le p \le n$ Processors

First, consider a hypercube with $n^2p,\ 1 \le p \le n$ processors. Such a hypercube can be viewed as an $n \times n \times p$ array. Let $PE(u, v, w)$ denote the processor in position (u, v, w), $0 \le u < n$, $0 \le v < n$, $0 \le w \le p$ of this array.

The $n \times n$ lattice graph is initially mapped onto the face $(u, v, 0)$ of the hypercube. This face is called face 0. $PE(u, v, 0)$ contains the weight of the (at most) three edges coming into vertex (u, v) of the lattice graph, $0 \le u, v < n$. Three registers: *left, diagonal,* and *up* are used for this purpose.

$$left(u, v, 0) = \begin{cases} 0 & v = 0 \\ weight((<u, v-1>, <u, v>) & v > 0 \end{cases}$$

$$diagonal(u, v, 0) = \begin{cases} 0 & u = 0 \text{ or } v = 0 \\ weight((<u-1, v-1>, <u, v>) & u > 0 \text{ and } v > 0 \end{cases}$$

$$up(u, v, 0) = \begin{cases} 0 & u = 0 \\ weight((<u-1, v>, <u, v>) & u > 0 \end{cases}$$

We shall say that processor $(u, v, 0)$ *represents* vertex (u, v) of the lattice graph.

9.4.1.1 Computing Boundary Distances

Since the computation of the boundary distances for all $k \times k$ subgraphs is done in parallel, we need consider only one of these subgraphs. Under the assumption that k is a power of 2, the processors

$$\{(u, v, w) \mid (u, v, 0) \text{ represents a vertex } (u, v) \text{ in the } k \times k \text{ subgraph}, \; 0 \le w < p\}$$

form a $k \times k \times p$ subhypercube.

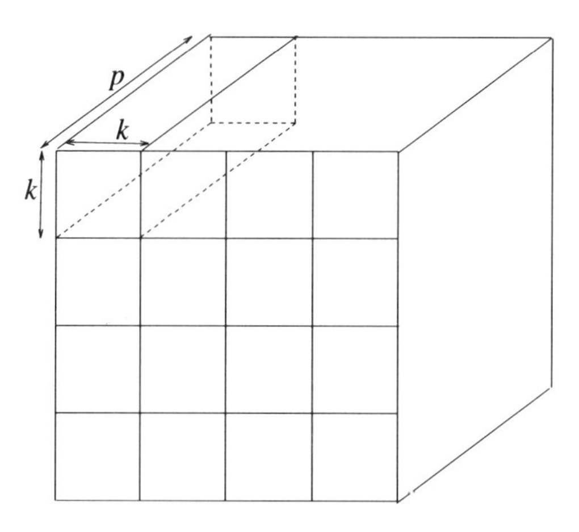

Figure 9.4 Decomposition into $k \times k \times p$ subhypercubes

The computation of the boundary distances for each $k \times k$ subgraph will be done by the corresponding $k \times k \times p$ subhypercube. To compute the boundary distances for any $a \times a$ subgraph of a $k \times k$ subgraph, the corresponding $a \times a \times p$ subhypercube will be used. Following this computation, the processors on face 0 of these $a \times a \times p$ subhypercubes will contain the boundary distances in register *TR*, *TB*, *LR*, and *LB*. Specifically, $XY(i, j, 0)$ will be the shortest distance from the i'th vertex on boundary X to the j'th vertex on boundary Y where $X \in \{T, L\}, Y \in \{R, B\}$, and i and j are relative to the respective $a \times a \times p$ subhypercube (i.e., the top left corner vertex in each such hypercube has $i = j = 0$). When computing for a $2a \times 2a$ subgraph, the initial configuration for face 0 of a $2a \times 2a$ subhypercube is shown in Figure 9.5 (a). I.e., the *TR*, *TB*, *LR*, and *LB* registers of face 0 of each $a \times a \times p$ subhypercube contain the corresponding boundary distances. Following the computation for the $2a \times 2a$ subgraph the boundary distances are to be distributed as in Figure 9.5. This will result in the correct initial condition for the computation of boundary distances for $4a \times 4a$ subgraphs.

We explicitly consider only the computation of the new *TR* values. The computation of the new *TB*, *LR*, and *LB* values is similar. The computation of the *TR* values is done in three stages. In the first stage the processors on face 0 of the top left $a \times a \times p$ subhypercube compute T_0R_2; on the top right subhypercube T_0B_1; and on the bottom right subhypercube T_1R_3 (Figure 9.6 (a)). The distances computed in the remaining two stages are shown in Figure 9.6 (b) and (c).

Stage 1 Computation

Equations (9.3) and (9.5) will be used to compute T_0R_2 and T_0B_1 respectively, The equation for T_1R_3 is:

$$T_1R_3(i, j) = \min\{ \min_{0 \le s < a} \{T_1B_1(i, s) + B_1T_3(s, s) + T_3R_3(s, j)\}, \quad (9.8)$$

$$\min_{0 \le s < a-1} \{T_1B_1(i, s) + B_1T_3(s, s+1) + T_3R_3(s+1, j)\}\}$$

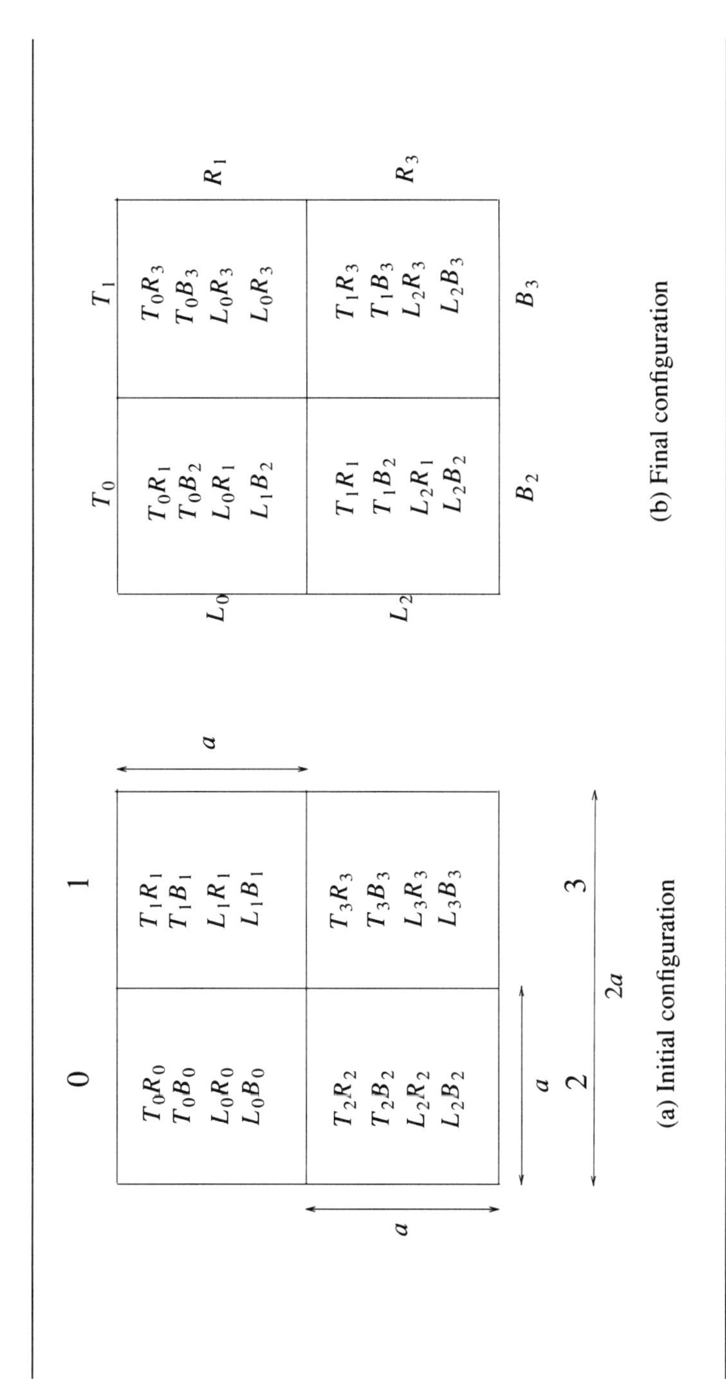

Figure 9.5 Initial/final configuration for $2a \times 2a$ subgraphs

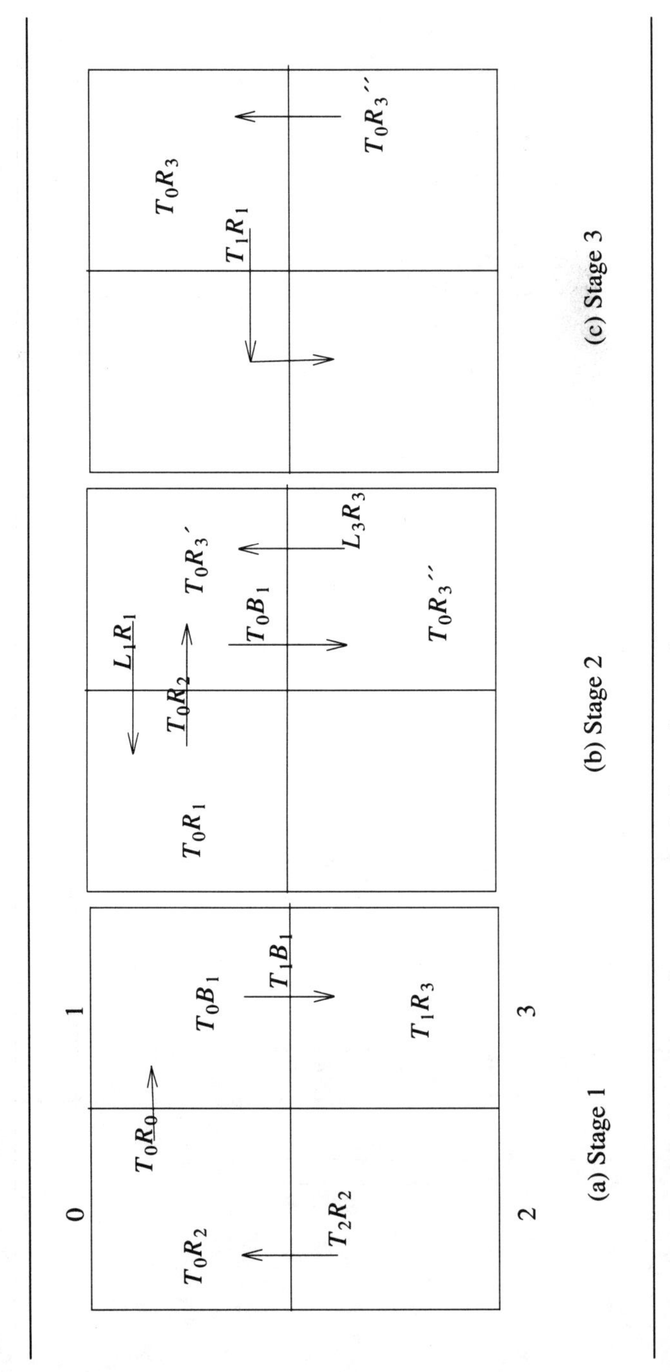

(a) Stage 1 (b) Stage 2 (c) Stage 3

Figure 9.6 Stages of boundary distance calculation of *TR* values

Equations (9.3), (9.5), and (9.8) may be rewritten into the form:

$$result(i, j) = \min\{E(i, j), F(i, j)\}$$

where $E(i, j)$ is the result of the $\min\limits_{0 \le s < a}$ part and $F(i, j)$ that of the $\min\limits_{0 \le s < a-1}$ part. The computation of $E(i, j)$ and $F(i, j)$ is very similar to the computation of the product, C, of two $a \times a$ matrices A and B. $C(i, j)$ is given by:

$$C(i, j) = \sum_{0 \le s < a} A(i, s) * B(s, j)$$

Replacing + by min and $*$ by +, we get

$$D(i, j) = \min_{0 \le s < a} \{A(i, s) + B(s, i)\} \qquad (9.9)$$

If in (9.8), we set $A(i, s) = T_1B_1(i, s)$ and $B(s, j) = B_1T_3(s, s) + T_3R_3(s, j)$ or $A(i, s) = T_1B_1(i, s) + B_1T_3(s, s)$ and $B(s, j) = T_3R_3(s, j)$, then $D(i, j) = E(i, j)$.

Let *MinSum* (A, B, D, a) be a hypercube procedure to compute D as in (9.9) in subhypercubes of size $a \times a \times p$. Such a procedure is easily obtained from the matrix multiplication procedure *MatrixMultiply* by using the above transformation. We assume that *MinSum* begins with $A(i, j)$ and $B(i, j)$ in processor (i, j) and leaves $D(i, j)$ in this processor when done.

The algorithm for the stage 1 computation is given in Program 9.1. The procedure *ColumnShift* (A, i, W) shifts the data in the columns of a hypercube down by i. For this purpose, the columns are divided into windows of size W. There is no wraparound and the fill is done using ∞'s. *RowShift* is analogous to *ColumnShift* except that it works on rows of a hypercube and does a leftward shift of i. Both of these procedures are simple adaptations of *SIMDShift*. The operator $\Leftarrow$ denotes a data broadcast.

In Step 1, we set up the A and B registers of the processors in squares 0, 1, and 3 (cf Figure 9.6(a)) so that a $MinSum(A, B, E, a)$ will result in $E(i, j)$ as above. For this, we need:

{square 0}
$A(i, j, 0) = T_0B_0(i, j) + B_0T_2(j, j)$
$B(i, j, 0) = T_2R_2(i, j)$

{square 1}
$A(i, j, 0) = T_0R_0(i, j)$
$B(i, j, 0) = R_0L_1(i, i) + L_1B_1(i, j)$

{square 3} $A(i, j, 0) = T_1B_1(i, j) + B_1T_3(j, j)$
$B(i, j, 0) = T_3R_3(i, j)$

Step 1: [Inititalize to compute $E(i, j)$ in register $E(i, j, 0)$]
{square 0}
$C^0(0, j, 0) \leftarrow up^2(0, j, 0), 0 \le j < a$
$C^0(i, j, 0) \Leftarrow C^0(0, j, 0), 0 \le i, j < a$
$A^0(i, j, 0) := TB^0(i, j, 0) + C^0(i, j, 0), 0 \le i, j < a$
$B^0(i, j, 0) \leftarrow TR^2(i, j, 0), 0 \le i, j < a$
{square 1}
$A^1(i, j, 0) \leftarrow TR^0(i, j, 0), 0 \le i, j < a$
$C^1(i, 0, 0) := left^1(i, 0, 0), 0 \le i < a$
$C^1(i, j, 0) \Leftarrow C^1(i, 0, 0), 0 \le i, j < a$
$B^1(i, j, 0) := C^1(i, j, 0) + LB^1(i, j, 0), 0 \le i, j < a$
{square 3}
$A^3(i, j, 0) \leftarrow TB^1(i, j, 0), 0 \le i, j < a$
$C^3(0, j, 0) := up^3(0, j, 0), 0 \le j < a$
$C^3(i, j, 0) \Leftarrow C^3(0, j, 0), 0 \le i, j < a$
$A^3(i, j, 0) := A^3(i, j, 0) + C^3(i, j, 0), 0 \le i, j < a$
$B^3(i, j, 0) := TR^3(i, j, 0), 0 \le i, j < a$

Step 2: [Compute E]
$MinSum(A, B, E, a)$

Step 3: [Initialize for F]
{square 0}
$C^0(0, j, 0) \leftarrow diagonal^2(0, j, 0)$
$RowShift(C^0, -1, a)$ {use ∞ fill}
$C^0(i, j, 0) \Leftarrow C^0(0, j, 0), 0 \le i, j < a$
$A^0(i, j, 0) := TB^0(i, j, 0) + C^0(i, j, 0), 0 \le i, j < a$
$ColumnShift(B^0, -1, a)$ {use ∞ fill}
{square 1}
$C^1(i, 0, 0) \leftarrow diagonal^1(i, 0, 0), 0 \le i < a$
$C^1(i, j, 0) \Leftarrow C^1(i, 0, 0), 0 \le i, j < a$
$B^1(i, j, 0) := C^1(i, j, 0) + LB^1(i, j, 0), 0 \le i, j < a$
$ColumnShift(B^0, -1, a)$ {use ∞ fill}
{square 3}
$A^3(i, j, 0) := A^3(i, j, 0) - C^3(i, j, 0), 0 \le i, j, < a$
$C^3(0, j, 0) \leftarrow diagonal^3(0, j, 0), 0 \le j < a$
$RowShift(C^3, -1, a)$ {use ∞ fill}
$A^3(i, j, 0) := A^3(i, j, 0) + C^3(i, j, 0), 0 \le i, j, < a$
$ColumnShift(B^0, -1, a)$ {use ∞ fill}

Step 4: [Compute F]
$MinSum(A, B, F, a)$

Step5: [Stage 1 Result]
$S1(i, j, 0) := \min\{E(i, j, 0), F(i, j, 0)\}, 0 \le i, j < a$

Program 9.1 Stage 1 Computation

The notation $X^q(i, j, 0)$ refers to register X of the processor in position $(i, j, 0)$ of square $q, 0 \le q \le 3$. So,

$$C^0(0, j, 0) \leftarrow up^2(0, j, 0)$$

denotes a data transfer from the *up* register of the processor in position $(0, j, 0)$ of square 2 to the C register of the processor in position $(0, j, 0)$ of square 0.

In Step 2, the E values are computed using *MinSum*. Step 3 sets up the A and B registers for the computation of F. This results in:

$$\{\text{square } 0\}$$
$$A(i,\ j,\ 0) = T_0B_0(i,\ j) + B_0T_2(j,\ j+1)$$
$$B(i,\ j,\ 0) = T_2R_2(i+1,\ j)$$

$$\{\text{square } 1\}$$
$$A(i,\ j,\ 0) = T_0R_0(i,\ j)$$
$$B(i,\ j,\ 0) = R_0L_1(i,\ i+1) + L_1B_1(i+1,\ j)$$

$$\{\text{square } 3\}$$
$$A(i,\ j,\ 0) = T_1B_1(i,\ j) + B_1T_3(j,\ j+1)$$
$$B(i,\ j,\ 0) = T_3R_3(i+1,\ j)$$

Note that the *RowShifts* of C^0 and C^3 can be done in parallel and the *ColumnShifts* of B^0, B^1, and B^3 can also be done in parallel. Steps 4 and 5 complete the Stage 1 computation.

Complexity of Stage 1

Steps 1 and 3 each take O($\log a$) time. This is due to the $\Leftarrow$ operations and the shifts. The time for Steps 2 and 4 is O($\frac{a}{z_a}$ + $\log z_a$), $z_a = \min\{a,p\}$. So, the total Stage 1 time is ($\frac{a}{z_a}$ + $\log a$), $z_a = \min\{a, p\}$.

Stage 2 Computation

This is very similar to the Stage 1 computation and can be completed in O($\frac{a}{z_a}$ + $\log a$) time.

Stage 3 Computation

T_1R_1 can be moved from square 1 to square 2 using two routes by following the path shown in Figure 9.6(c). The data movements for the computation of T_0R_3 take $O(\log a)$ time.

Overall Complexity

The overall time needed to compute boundary distances for $a \times a$ subgraphs is $O(\frac{a}{z_a} + \log a)$, $z_a = \min\{a,p\}$. The time to compute the boundary distances for all $k \times k$ subgraphs is therefore

$$O(\sum_{a=2,4,..,k} (\frac{a}{z_a} + \log a)) = O(\frac{k}{p} + \log^2 k)$$

9.4.1.2 Computing $cost(n-1, n-1)$

The remaining computations needed to obtain $cost(n-1, n-1)$ can be performed in $O(\frac{n}{k}\log k)$ time using only those processors that are on face 0 of the hypercube. First, the $k \times k$ subhypercubes of face 0 compute the LL and TT values. The processor in position $(l, j, 0)$ of a $k \times k$ hypercube computes $LL(l, j)$ and $TT(l, j)$. We describe the computation for TT only. The computation of LL is similar.

Computing TT

From Section 9.3 and the definition of $left(i, j, 0)$, we see that $TT(l, j, 0) = \sum_{r=l}^{j-1} left(0, r+1, 0) = \sum_{r=l+1}^{j} left(0, r, 0)$. Hence, the TT values may be computed in face 0 using the strategy of Program 9.2. The computation of TT is viewed as several prefix sum computations; one for each value of l. Steps 1 and 2 set up each row of the $k \times k$ subhypercube so that a prefix sum of the $T1$ values on that row will result in the correct TT value. Procedure *PrefixSum* is a suitably modified version of procedure

SIMDPrefixSum. Steps 1 and 3 take O(log$k$) time and Step 2 takes O(1) time.

Step 1: [Broadcast *left* (0, r, 0) over columns]
$T1(0, r, 0) := left(0, r, 0),\ 0 \le r < k$
$T1(s, r, 0) \Leftarrow T1(0, r, 0),\ 0 \le s < k$

Step 2: [Zero out values not needed]
$T1(l, r, 0) := 0,\ 0 \le r \le l < k$

Step 3: [Compute $TT(l, j, 0)$]
PrefixSum (TT, k, $T1$) {Prefix sum on rows}

Step4: [Clean up]
$TT(l, j, 0) := \infty,\ 0 \le j < l < k$

Program 9.2 Computing $TT(l, j, 0)$

Once the TT and LL values have been computed, we proceed to compute the L, T, B, and R values defined in Section 9.3. While all the face 0 processors of a $k \times k \times p$ subhypercube are involved in the computation of the L, T, B, and R values for that subhypercube, the final values are stored only in certain processors. The assignment is

$$\left.\begin{array}{l} L(j, 0, 0) = L_i(j) \\ T(0, j, 0) = T_i(j) \\ R(j, k-1, 0) = R_i(j) \\ B(k-1, j, 0) = B_i(j) \end{array}\right\} \quad 0 \le j < k$$

where i is the label of the corresponding $k \times k$ subgraph (Figure 9.3). Since the ideas used in the computation of L, T, R, and B are quite similar, we describe only the computation of one part of $L_b(j)$. Specifically, we consider computing

$$L'_b(j) = \min_{0 \le s \le j} \{R_a(s) + R_aL_b(s, s) + L_bL_b(s, j)\}$$

The value $L'_b(j)$ will be left in register $LX(j, 0, 0)$ of the $k \times k$ subhypercube that represents subgraph b. The strategy to compute L'_b is given in Program 9.3. Following Step 1, we have $R(s, 0, 0) = R_a(s)$, $0 \le s < k$. Following Step 2, $R(s, j, 0) = R_a(s) + R_a L_b(s, s)$. Note that $R(s, j, 0) + LL(s, j, 0)$ is a term in the *min* for $L'_b(j)$. To compute $L'_b(j)$ we need simply find the minimum of all the $R(s, j, 0) + LL(s, j, 0)$ values in column j. This is true as $LL(s, j) = \infty$ for $s > j$. Step 3 computes this minimum in $LX(0, j, 0)$. Step 4 routes the LX values to the proper processors.

Step 1: [Bring in R_a values]
Shift in the R values in the $1 \times k$ column of processors to the left of this $k \times k$ subhypercube right by 1. Put the values in the R registers of these processors.

Step 2: [Add with *left* and row broadcast]
$R(s, 0, 0) := R(s, 0, 0) + left(s, 0, 0)$, $0 \le s < k$
Broadcast $R(s, 0, 0)$ to $R(s, j, 0)$, $0 \le s, j < k$

Step 3: [Compute $L'_b(j)$ in processor $(0, j, 0)$]
$LX(0, j, 0) := \min_{0 \le s < j} \{R(s, j, 0) + LL(s, j, 0)\}$

Step 4: [Route to correct processors]
Route $LX(0, j, 0)$, to $LX(j, 0, 0)$, $0 \le j < k$

Program 9.3 Computing LX

Step 1 is a shift of 1 in a window of size $2k$. This takes $O(\log k)$ time. Steps 2 and 3 are easily seen to take $O(\log k)$ time. Step 4 is a BPC permutation of LX values and can also be done in $O(\log k)$ time. Since the L, B, T, R values of all $k \times k$ subgraphs are computed in $O(n/k)$ steps with each step computing these values for some of the $k \times k$ subgraphs, the total time taken to compute the L, B, T, R values for all $k \times k$ subgraphs is $O(\frac{n}{k}\log k)$.

Combining this with the time required to compute the boundary distances, we get $O(\frac{k}{p} + \log^2 k + \frac{n}{k}\log k)$ as the time taken to compute $cost(n-1, n-1)$ beginning with the initial input data. This is minimum when k is approximately $\sqrt{np\log n}$. Substituting this for k, we get $O(\sqrt{\frac{n\log n}{p}} + \log^2 n)$ as the complexity of our algorithm to compute $cost(n-1, n-1)$, The number of processors used is $n^2 p$, $1 \le p \le n$.

9.4.1.3 Traceback

When $p \ge \log k = \log(\sqrt{np\log n})$, O(1) memory per processor is enough to remember the $O(n^2 \log k)$ decisions made during the computation of the boundary distances and the further $O(n^2/k)$ decisions made to compute $cost(n-1, n-1)$ from the boundary distances. The entry/exit points for all the subgraphs can be computed in $O(\frac{n}{k}\log k)$ time using the latter information. The paths within each $k \times k$ subgraph can be constructed in $O(\log^2 k)$ time. So, the traceback takes $O(\frac{n}{k}\log k + \log^2 k) = O(\sqrt{\frac{n\log n}{p}} + \log^2 n)$ when $k = \sqrt{np\log n}$.

When $p < logk$, we do not have enough memory to save the decisions made during the computation of the boundary distance. However, we do have enough memory to save the later $O(n^2/k)$ decisions. Hence, the entry and exit vertices of the $k \times k$ subgraphs can still be determined in $O(\frac{n}{k}\log k)$ time. The shortest path from an entry vertex to an exit vertex of a $k \times k$ subgraph may be found by first modifying the edge costs of each $k \times k$ subgraph so that the shortest (0, 0) to $(k-1, k-1)$ path will contain the shortest entry to exit path. Figure 9.7 shows one of the possible cases for the entry and exit vertices. Edges on the path from (0, 0) to the entry vertex are given a cost of 0. Also those on the path from the exit vertex to $(k-1, k-1)$ are given a cost of 0. Remaining edges not contained in the rectangle defined by the entry and exit vertices are given a cost of ∞. As a result of this transformation, finding a shortest entry to

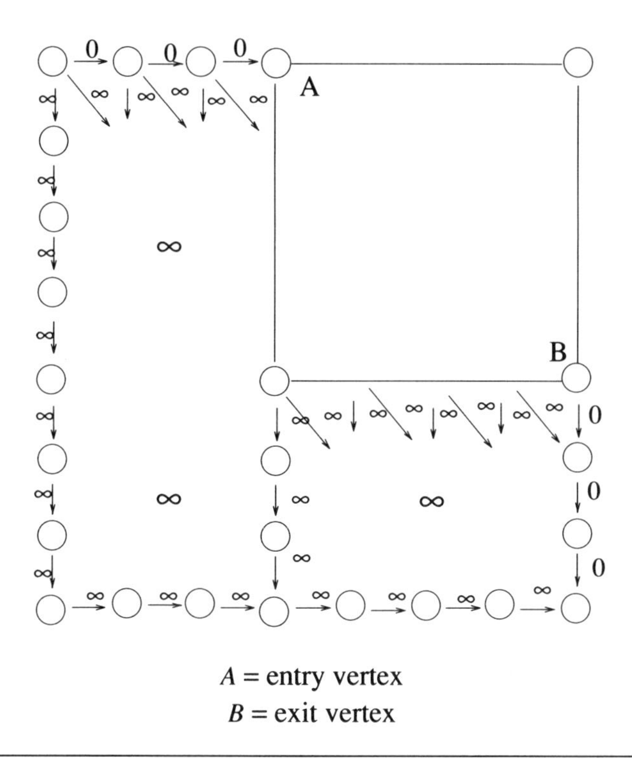

Figure 9.7 Modifying a $k \times k$ subgraph

exit path is equivalent to finding a shortest (0, 0) to $(k-1, k-1)$ path. This can be done by recursive application of the algorithm on the $k \times k$ subgraphs. So, we compute new boundary distances, new $cost(k-1, k-1)$ values, etc.

The run time $T(n, p)$ of the trace back when $p < \log k = \log(\sqrt{np \log n})$ is

$$T(n, p) = O(\sqrt{\frac{n \log n}{p}}) + T(\sqrt{np \log n}, p)$$

$$= O\left[\sum_i \left[\frac{n \log n}{p}\right]^{\frac{1}{2^i}}\right] = O\left[\sqrt{\frac{n \log n}{p}}\right]$$

So, regardless of whether $p \le \log k$ or $p \ge \log k$, the traceback can be completed in $O\left(\sqrt{\frac{n\log n}{p}} + \log^2 n\right)$ time using n^2p, $1 \le p \le n$ processors and O(1) memory per processor.

9.4.2 p^2, $n\log n \le p^2 < n^2$ Processors

The algorithm for this case may be obtained from that for the case of n^2p, $1 \le p \le n$ processors by first setting p to 1 to get the algorithm for the case of n^2 processors. This simply requires eliminating the third dimension in the earlier algorithm and replacing the *MinSum* procedure with an equivalent procedure for n^2 processors. Next, we use the ideas of Chapter 3 to go from an n^2 processor algorithm to a p^2 processor algorithm ($1 \le p \le n$). This requires us to map each $\frac{n}{p} \times \frac{n}{p}$ subgraph of the $n \times n$ lattice graph onto a single processor. Hence, each processor of the $p \times p$ hypercube contains the information corresponding to an $\frac{n}{p} \times \frac{n}{p}$ subgraph. The algorithm to compute $cost(n-1, n-1)$ takes the form given in Program 9.4.

Step 1: Each processor computes the boundary distances for its $\frac{n}{p} \times \frac{n}{p}$ subgraph.

Step 2: **for** $a := 1, 2, 4, ..., k/2$ **do**
  Each $2a \times 2a$ subhypercube computes the boundary distances for its vertices from those for its constituent $a \times a$ subhypercubes.

Step 3: Combine the boundary distances of $k \times k$ subhypercubes.

Program 9.4 Algorithm for p^2, $n\log n \le p^2 \le n^2$ processors

Step 1 is done using the serial dynamic programming algorithm for string editing in each processor. The dynamic programming algorithm is used $2n/p - 1$ times; once for each of the $2n/p - 1$ vertices in the top and left boundaries of the $n/p \times n/p$ subgraph. So, in each application the shortest distances from one of these $2n/p - 1$ vertices to all $2n/p - 1$ vertices in the right and bottom boundaries are found. Since each application of the dynamic programming algorithm takes $O(n^2/p^2)$ time, the total time for Step 1 is $O(n^3/p^3)$.

Step 2 is done using a modified version of the n^2 processor algorithm. This modification is similar to that described in Chapter 3 for matrix multiplication. The time taken is $O\left(\left(\frac{n}{p}\right)^3 \sum a\right) = O\left(k\left(\frac{n}{p}\right)^3\right)$.

The combining of the boundary distances of the $k \times k$ subgraphs is done in $O(p/k)$ iterations. Note that $k \times k$ subhypercubes form a $p/k \times p/k$ hypercube. In each iteration of the $O(p/k)$ iterations needed to compute $cost(n-1, n-1)$ at most one $k \times k$ subhypercube of each column of Figure 9.8 is active. When $p^2 < n^2$, we can put the remaining processors in each column to use. If $\frac{n^2}{p^2} \le \frac{p}{k}$, then we may group the processors in each column such that each group contains $\frac{n^2}{p^2}$ $k \times k$ subhypercubes from the same column. In Figure 9.8, $n^2/p^2 = 4$ and $p/k = 8$. The n^2/p^2 $k \times k$ subhypercubes in each group may themselves be viewed as an $n/p \times n/p$ array (2×2 array in Figure 9.8) which is drawn as a one dimensional array of size n^2/p^2. The pairs (i, j) outside Figure 9.8 give such an interpretation for the top left group. Hence, we may refer to a processor using the tuple

$$[a, b, c, d, e, f]$$

where (a, b) indexes the processor group, $0 \le a < p^3/n^2$, $0 \le b < p/k$; (c, d) indexes a $k \times k$ subhypercube within a group, $0 \le c, d < n/p$; and (e, f) indexes a processor within a $k \times k$ subhypercube, $0 \le e, f < k$. For processors in the top left $k \times k$ subhypercube of Figure 9.8, $a = b = c = d = 0$. For those in the $k \times k$ subhypercube below this one, $a = b = c = 0, d = 1$. For

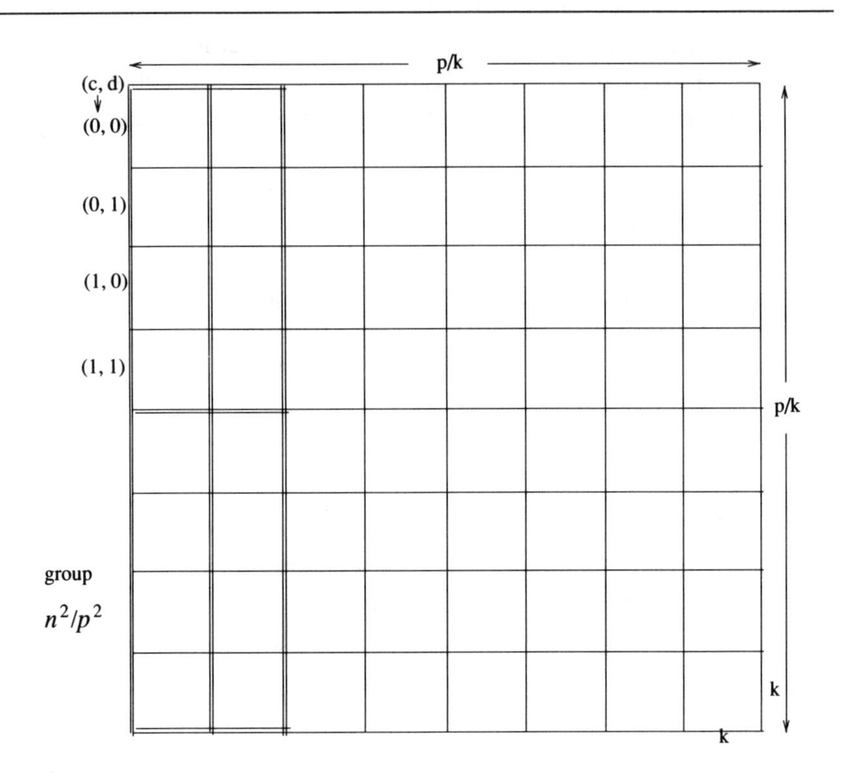

Figure 9.8 p^2 processors, $p/k = 8$, $n^2/p^2 = 4$

the bottom left $k \times k$ subhypercube of Figure 9.8, $a = 1$, $b = 0$, $c = d = 1$. Each processor in a $k \times k$ subhypercube represents $\frac{n^2}{p^2}$ (or an $n/p \times n/p$ subgraph) vertices of the original $n \times n$ lattice graph. Let (g, h), $0 \le g, h < n/p$ index these vertices. We may use the tuple

$$[a, b, c, d, e, f, g, h]$$

to refer to vertex (g, h) of the (e, f) processor in the $k \times k$ subhypercube (c, d) of group (a, b). In the algorithms of Section 1.1 each vertex was represented by a face 0 processor. Now, each processor contains distances for n^2/p^2 of these face 0 processors. Let $dist[a, b, c, d, e, f, g, h]$ denote these distances for processor (g, h) of the (e, f) processor in the $k \times k$ subhypercube (c, d) of group (a, b) of face 0. Following step 2 of Program 9.4, $dist[a, b, c, d, e, f, g, h]$ is in processor $[a, b, c, d, e, f]$. Using the BPC routing algorithm of Chapter 2 we can, in $O\left(\frac{n^2}{p^2}\log\frac{n^2k^2}{p^2}\right)$ time rearrange $dist$ such that

$$dist'[a, b, g, h, e, f, c, d] = dist[a, b, c, d, e, f, g, h]$$

Now, the n^2/p^2 $dist$ values formerly in a single processor have been distributed to the $\frac{n^2}{p^2}$ corresponding processors in the same group. As a result of this distribution, we can essentailly use the algorithm of Section 9.4.2 to combine boundary distances in $O(\log n)$ time per iteration. So the total Step 3 time becomes $O\left(\frac{n^2}{p^2}\log\frac{n^2k^2}{p^2} + \frac{p}{k}\log n\right)$ = $O\left(\frac{n^2}{p^2}\log n + \frac{p}{k}\log n\right)$ (note that $n^2k^2/p^2 < p^2 < n^2$). The overall time for Program 9.4 is therefore $O(k(n/p)^3 + \frac{n^2}{p^2}\log n + \frac{p}{k}\log n)$. If we set $k = (p^2\sqrt{\log n})/n^{3/2}$, then since $\sqrt{n}\log n \le p$, $n^2/p^2 \le p/k$. Further, with this k the run time becomes $O\left(\frac{n^{3/2}}{p}\sqrt{\log n}\right)$.

The traceback needed to obtain the shortest $(0, 0)$ to $(n-1, n-1)$ path can be done in the above time by either using $O\left(\frac{n^2}{p^2}\log\frac{p^2}{n}\right)$ memory per processor to store the decisions made during steps 1, 2, and 3 of Figure 12 or by using $O\left(\frac{n^2}{p^2}\right)$ memory per processor to save only the decisions made in Steps 1 and 3 and recomputing those for Step 2.

The complexities of our algorithms for n^2p, $1 \le p \le n$ and p^2, $n\log n \le p^2 < n^2$ processors may be restated as $O\left(n\sqrt{\frac{\log n}{l}} + \log^2 n\right)$ for nl, $\log n \le l \le n^2$ processors.

References

1. S. B. Akers, D. Harel and B. Krishnamurthy (1987), "The star graph: An attractive alternative to the n-Cube", *Proc. of Intl. Conference on Parallel Processing.*
2. W. C. Athos and C. L. Seitz (1988), "Multicomputers: Message-passing concurrent computers", *IEEE Computer*, 21, 8, pp. 9-25.
3. D. H. Ballard and C. M. Brown (1985), *Computer Vision*, Prentice Hall, New Jersey.
4. T. F. Chan and Y. Saad (1986a), "Multigrid algorithms on the hypercube multiprocessor", *IEEE Transactions on Computers*, C-35, Nov.
5. M. Y. Chan (1986b), "Dilation-2 embeddings of grids into hypercubes," *Proceedings, 1988 International Conference on Parallel Processing*, Vol 3, pp 295-297.
6. R. E. Cypher, J. L. C. Sanz (1987), "The Hough transform has $O(N)$ complexity on SIMD $N \times N$ Mesh Array Architectures," *Proc. of IEEE CAPAMI Workshop.*
7. E. Dekel, D. Nassimi and S. Sahni (1981), "Parallel matrix and graph algorithms", *SIAM Jr. on Comp.*, 10, 4, pp. 657-675.
8. E. Dekel and S. Sahni (1983), "Binary trees and parallel scheduling algorithms", *IEEE Transactions on Computers*, Vol. C-32, No. 3, March, pp. 307-315.
9. R. O. Duda and P. E. Hart (1973), *Pattern Classification and Scene Analysis*, John Wiley and Sons.
10. Z. Fang, X. Li and L. M. Ni (1985), "Parallel algorithms for image template matching on hypercube SIMD computers", *IEEE CAPAMI Workshop*, pp 33-40.
11. Z. Fang and L. M. Ni (1986), "Parallel algorithms for 2-D convolution", *International Conference on Parallel Processing*, pp 262-269.

11. M. J. Flynn (1966), "Very high-speed computing systems", *Proceedings of the IEEE,* 54 , Dec., pp. 1901-1909.

12. K. S. Fu (1974), *Syntactic Methods in Pattern Recognition*, Academic Press.

13. K. Fukunaga (1972), *Introduction to Statistical Pattern Recognition*, Academic Press.

14. J. Gustafson (1988), "Reevaluating Amdahl's Law," CACM, 31, 5, pp. 532-533.

15. E. Horowitz and S. Sahni (1987), *Fundamentals of data structures in Pascal,,* Second Edition, Computer Science Press, Maryland.

16. O. H. Ibarra, T. C. Pong and S. Sohn (1988), "Hypercube algorithms for some string comparision problems", *University of Minnesota Technical Report*, Jan. A short version of this appears in *IEEE Trans. On ASSP*, 1989.

17. S. L. Johnsson (1987), "Communication efficient basic linear algebra operations on hypercube architectures", *Jr of Parallel and Distributed Computing*, 4, 2, pp 133-172.

18. D. Knuth (1973), *The art of computer programming: Sorting and searching*, Vol 3, Addison Wesley, NY.

19. V. Prasanna Kumar, and V. Krishnan (1987), "Efficient template matching on SIMD arrays," *1987 International Conference on Parallel Processing*, The Pennsylvania State University Press, pp. 765-771.

20. V. Kumar, V. Nageshwara, and K. Ramesh (1988), "Parallel depth first search on the ring architecture," *Proceedings of 1988 International Conference on Parallel Processing*, pp. 128-132, Penn State University Press.

21. T. Lai, and S. Sahni (1984), "Anomalies in parallel branch and bound algorithms," *Communications of ACM*, Vol. 27, pp. 594-602.

22. S. Y. Lee, S. Yalamanchali and J. K. Agarwal (1987), "Parallel image normalization on a mesh connected array processor", *Pattern Recognition*, Vol. 20, No. 1, Jan, pp. 115-120.

23. X. Li and Z. Fang (1986), "Parallel Algorithms for clustering on hypercube SIMD computers", *Proceedings of 1986 Conference on Computer Vision and Pattern Recognition*, pp. 130-133.

24. G. Li, and B. Wah (1986), "Coping with anomalies in parallel branch-and-bound algorithms," *IEEE Trans. on Computers*, No. 6, C-35 (June), pp. 568-572.

25. H. Liu and K. Fu (1985), "VLSI arrays for minimum-distance classifications," *VLSI for pattern recognition and image processing*, Springer-Verlag.

26. D. Nassimi, and S. Sahni (1981), "Data Broadcasting in SIMD computers," *IEEE Transactions on Computers*, No. 2, Vol. C-30 (Feb.), pp. 101-107

27. D. Nassimi and S. Sahni (1982), "Optimal BPC permutations on a cube connected computer", *IEEE Transactions on Computers*, No. 4, Vol. C-31, April, pp. 338-341.

28. M. Quinn (1987), *Designing efficient algorithms for parallel computers*, McGraw-Hill Inc.

29. M. Quinn and N. Deo (1986), "An upper bound for the speedup of parallel branch-and-bound algorithms," BIT, 26, No. 1 (March), pp. 35-43.

30. S. Ranka and S. Sahni (1988a), "Image template matching on MIMD hypercube multicomputers," *Proceedings 1988 International Conference on Parallel Processing*, The Pennsylvania State University Press, pp. 92-99.

31. S. Ranka and S. Sahni (1988b), "Image template matching on SIMD hypercube multicomputers," *Proceedings 1988 International Conference on Parallel Processing*, The Pennsylvania State University Press, pp. 84-91.

32. S. Ranka and S. Sahni (1988c), "String editing on an SIMD hypercube multicomputer," University of Minnesota Technical Report.

33. S. Ranka and S. Sahni (1988d), "Clustering on a hypercube multicomputer," Technical Report, University of Minnesota.

34. S. Ranka and S. Sahni (1989a), "Computing Hough transforms on hypercube multicomputers,'' Technical Report 89-1, University of Minnesota.

35. S. Ranka and S. Sahni (1989b), "Hypercube algorithms for image transformations," *Proceedings 1989 International Conference on Parallel Processing*, The Pennsylvania State University Press, Vol III, pp III-24 through III-31.

36. A. P. Reeves and C. H. Francfort (1985), "Data Mapping and Rotation Functions for the Massively Parallel Processor", *IEEE Computer Society Workshop on Computer Architecture for Pattern Analysis and Image Database Management*, pp. 412-417.

37. A. Rosenfeld (1987), "A note on shrinking and expanding operations on pyramids", *Pattern Recognition Letters 6,* Sept, pp. 241-244.

38. A. Rosenfeld and A. C. Kak (1982), *Digital Picture Processing*, Academic Press.

39. H. Siegel (1979), "Interconnection networks for SIMD machines", *IEEE Trans. on Computers,* 12, June, pp. 57-65.

40. Q. F. Stout (1987), "Supporting divide-and-conquer algorithms for image processing", *Jr of Parallel and Distributed Computing*, Vol 4, pp 95-115.

41. J. T. Tou and R. C. Gonzalez (1974), *Pattern Recognition Principles*, Addison-Wesley.

42. R. A. Wagner and M. J. Fischer (1974), "The string-to-string correction problem", *JACM*, Vol. 21, No. 1, Jan, pp. 168-173.

43. A. Wu (1985), "Embedding of tree networks into hypercubes," *Jr of Parallel and Distributed Computing*, 2, 3, pp 238-249.

INDEX